LASERS
AND THEIR
APPLICATIONS

LASERS
AND THEIR
APPLICATIONS

M. J. BEESLEY
B.SC., M.SC., M.INST.P.

Royal Aircraft Establishment
Ministry of Defence

TAYLOR & FRANCIS LTD
10–14 Macklin Street London WC2B 5NF

1976

First published 1971 by Taylor & Francis Ltd
10–14 Macklin Street, London WC2B 5NF

Reprinted 1972 (with minor corrections)
2nd Edition published 1976

ISBN 0 85066 045 9

Printed and bound in Great Britain by
Taylor and Francis (Printers) Ltd,
Rankine Road, Basingstoke,
Hampshire RG24 0PR.

Distributed in the United States of America
and its territories by Halsted Press (a division
of John Wiley & Sons Inc.) 605, Third Avenue,
New York, N.Y. 10016

Acknowledgments

I should like to thank the many former colleagues at the Services Electronic Research Laboratory who have given advice, help and information. In particular I am indebted to Mr. H. Foster for his help with the section on ring lasers and to Dr. R. Taylor and Mr. W. Barr for their many useful comments.

My gratitude is also due to Mr. P. Shilham whose work on the manuscript has been invaluable.

Acknowledgment should be expressed to the many firms and organizations which have supplied data and photographs and to various journals. All these sources have been indicated at the appropriate points throughout the book.

This book is published by kind permission of the Ministry of Defence (Navy Dept.).

Introduction

The aim of this book is to explain the mechanism of laser action, to describe the various types of laser which are commercially available and to describe some applications of the laser, both immediate and potential.

The invention of the laser in 1960 has provided scientists with a new and powerful tool which has many practical applications, as well as being a valuable aid to fundamental research. The applications of lasers cover a wide range from medicine to civil engineering and from chemical analysis to communications. It is apparent therefore that many practising scientists and engineers, as well as students, will need to obtain an understanding not only of basic principles but also of the construction and use of lasers. In this way the limitations as well as potentials will be appreciated.

It is hoped that this book goes some way towards answering this need.

Contents

1.1. *The Nature of Light*

IN order to understand how a laser works it will be useful to recount how present ideas concerning the nature of light have come about. Of all man's senses that of vision is the most important. The connection between vision and the rising and the setting of the sun was probably one of the first scientific deductions made by man. The Greeks were almost certainly the first to try to explain how vision takes place and consequently to conjecture on the nature of light itself. They had two theories : first that the eye reached out by means of probes or tentacles and touched the object which was thus ' seen '. Secondly that the object itself emitted some sort of material which was collected by the eye to give the sensation of vision. These theories are referred to as the tactile and emission theories.

The advent of experimental science in the 17th century brought about the abandonment of the tactile theory and the development of two emission theories each passionately defended by their respective proponents. These two emission theories were the corpuscular theory of Isaac Newton and the wave theory of Robert Hooke and Christian Huygens.

Even at that time a considerable amount of experimental data was available. Reflection and refraction had been known from antiquity. Interference (although it was not referred to as such until Thomas Young propounded his theories in the early 19th century) had been observed independently by Hooke and Robert Boyle in 1665 in the form of the somewhat unfairly named Newton's rings. In the same year diffraction had been observed by Grimaldi. Four years later Bartholinus discovered double refraction in a crystalline material called Iceland Spar.*

The wave theory, initially proposed in a primitive form by Hooke, then advanced and refined by Huygens, accounted very well for reflection, refraction and double refraction and the latter led Huygens to the conclusion that in some way light could be polarized in different directions. Huygens based his explanations on the principle that every point on a wavefront acted as a secondary source of spherical waves which were propagated through an all pervading medium called the aether. At that time transverse waves, where the direction of vibration

* For a full account of these terms and others mentioned in this section the reader is advised to consult one of the standard textbooks on light[1,2].

1

is at right angles to the direction of propagation, were well known in the form of water waves. Also familiar as sound waves were longitudinal waves. In this type of wave the direction of vibration is parallel to the direction of propagation. That light waves were longitudinal in nature was universally accepted by the supporters of the wave theory. This seems an odd choice as variation in polarization can be explained elegantly on a transverse wave theory but is quite inexplicable in the case of longitudinal waves. It may be that the association of sound and vision led to the idea that the waves in each case must be of the same type.

This inability to account for variations in polarization on a longitudinal wave theory led Newton to devise a corpuscular explanation. Newton thought of light as particles obeying his dynamical laws of motion and accounted for rectilinear propagation by assuming that the light particles had no mass and so, according to his laws of motion, could not be changed from their straight line trajectories by any impressed forces. Newton thought that rectilinear propagation was inconsistent with a wave theory as a light wave would be expected to diffract, i.e., to spread round corners and he devised ingenious explanations to account for Grimaldi's earlier observations. Newton was, of course, quite correct in thinking that light waves must exhibit diffraction but he failed to realise that if the wavelength is small enough any such diffraction would be extremely difficult to observe.

For a century or so the might of Newton's authority held sway and the wave theory did not obtain general acceptance, albeit with some notable exceptions, particularly that of the mathematician Euler, who correctly associated waves of different frequencies with different colours.

In 1801 Thomas Young in his classical two-slit experiment showed that light from two sources could combine to form regions of brightness and darkness called fringes. These could only be explained in terms of a wave theory, a bright fringe being formed where the two waves combine in phase so as to reinforce one another; and dark fringes being formed where the two waves find themselves out of phase and hence cancelling each other. Young termed these phenomena constructive and destructive interference respectively.

The colours seen when thin films are illuminated with white light were also explained by Young in terms of a wave theory where, like Euler, he associated different colours with different wavelengths. At about the same time Malus discovered polarization by reflection and in 1816 Fresnel and Arago showed that two waves polarized at right angles could not interfere. Young suggested that the only possible explanation for these observations was that light must not only be a wave but that, in addition, it had to be of a transverse nature.

Additional evidence in favour of the wave theory was provided by Fresnel, who explained diffraction quantitatively on the wave theory. Further conclusive proof was obtained in 1850, when Foucault measured the speed of light in air and water. According to the corpuscular theory

2

the denser material would attract the corpuscles hence speeding up the light, whereas on the wave theory the converse would be true. Foucault found that the speed of light in air was faster than in water and thus disproved the corpuscular theory.

A dramatic advance in the understanding of light resulted from the work on electricity and magnetism by Faraday, Oersted and Henry. In 1864 James Clerk Maxwell combined all the experimental data into a set of equations. From this set of equations could be deduced the existence of a wave with the property that its speed, c, in free space bore a simple relationship to the dielectric constant (permittivity), ε_0 and the magnetic permeability μ_0. This relationship is expressed in equation 1.1:

$$c = \sqrt{\frac{1}{\mu_0\, \varepsilon_0}} \tag{1.1}$$

Now the extraordinary property of this wave was that on substituting known values for μ_0 and ε_0 in equation 1.1 a result identical to the velocity of light in a vacuum was obtained, the latter having been previously found by completely different methods. Maxwell therefore proposed that light was an example of what he termed electromagnetic radiation having approximately a speed of 3×10^8 ms^{-1}, a frequency of 5×10^{14} Hz and a wavelength of $0{\cdot}5 \times 10^{-6}$ m. In its plane polarized form electromagnetic radiation and hence light is thought of as consisting of electric and magnetic vectors oscillating at right angles to each other and to the direction of propagation.

By discharging an induction coil across a spark gap and so setting up oscillating electric and magnetic fields, Hertz succeeded in generating non-visible electromagnetic waves of wavelength 10 m and frequency 5×10^7 Hz. He showed that these waves could be reflected and refracted in exactly the same way as light waves. Light in fact is just one small region of the electromagnetic spectrum which stretches from the long radio waves through the visible out to γ rays and cosmic rays. Figure 1.1 shows the range of electromagnetic waves with their frequency and wavelength.

Hertz can be thought of as starting the quest which has engaged a multitude of scientists for over a century: that is to make devices which will push the frequencies available 'artificially' higher and higher. The laser marks the present frontier at about 10^{15} Hz.

The concept of the aether was finally abandoned as it was realised that a material medium was not necessary for the propagation of electromagnetic transverse waves and the continual postulation of an elastic medium, in which light had to propagate, brought more problems than supplied answers.

The theories of Maxwell had difficulty in gaining acceptance particularly as it was not easy to envisage any mechanical model of electromagnetic radiation.

3

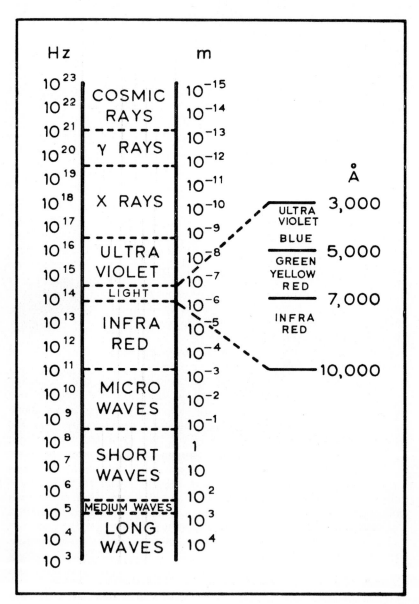

Fig. 1.1. The electromagnetic spectrum.

Despite the elegant way in which Maxwell's theory united several different branches of physics there were still some questions left unanswered. While the electromagnetic theory accounted so well for reflection, refraction, polarization, etc., it failed completely in trying to explain emission and absorption.

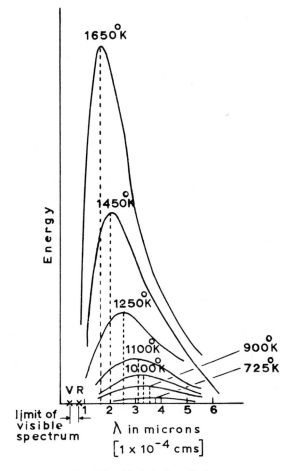

Fig. 1.2. Black-body radiation.

The emission of radiation had been studied by many physicists and it was concluded that when a perfectly emitting object, known as a black body, is in equilibrium with its surroundings, it has a characteristic spectral distribution which is shown in fig. 1.2. The distribution of frequencies and the total power emitted changes with temperature.

5

Classical analysis of black body radiation led to results which were quite different from those indicated in fig. 1.2. For instance on a classical basis the intensity of radiation should increase as the wavelength becomes shorter ; this so called 'ultraviolet catastrophe' was quite contrary to the observed evidence. However electromagnetic waves can be thought of as being produced by oscillating electric fields. Max Planck showed that a satisfactory explanation of black-body radiation could be obtained if the energy of the oscillators were allowed to take only certain values, namely integral values of hv where h is a constant called Planck's constant and v is the frequency of the electromagnetic wave produced by the oscillator. It should be noted that this suggestion by Planck, although revolutionary, was still consistent with a wave theory.

In the work of Hertz mentioned previously it was noticed that the spark could be made to pass more easily if the gap between the metal electrodes was illuminated with ultraviolet light. This phenomenon was accounted for in 1899 by J. J. Thomson who showed that the ultra violet radiation caused electrons, discovered by him in 1895, to be emitted from the surface of the metal so facilitating the passage of an electric current. This process is called the photoelectric effect.

The quantum theory of Planck was taken a step further by Einstein in 1905 who explained some observations of Lenard on the photoelectric effect. Lenard noted two peculiar features: first, if the frequency of the incident light was lowered sufficiently no electrons were emitted however intense the radiation. Secondly, electrons were emitted almost instantaneously when the light was allowed to fall on the metal irrespective of the intensity of the light, providing of course, that it was sufficiently high. On a wave theory of light this would be unexpected as the energy in a wave is spread out uniformly across its wavefront. With an extremely small particle, such as an electron, some time might be expected to elapse after the illumination is switched on before the electron receives sufficient energy to cause its ejection, particularly if the incident intensity were very low.

Einstein concluded from these observations that only one explanation was possible: that electromagnetic radiation could be described in the form of particles of energy which he named photons. The energy E of a photon is proportional to the frequency of the radiation v or

$$E = hv \tag{1.2}$$

the constant of proportionality being in fact Planck's constant, h.

Thus in 250 years the theories of light had turned full circle almost back to the ideas of Newton. With the important exception, however, that, depending on the nature of the experiment, light behaved as either a wave or a particle. The wave theory of light accounts for effects like interference and diffraction which cannot possibly be explained on a particle theory. On the other hand the wave theory fails with black-body radiation and the photoelectric effect which require a particle theory. In the case of low frequency electromagnetic radiation, such as radio waves, the energy of each quantum is very small and so even a small amount of such energy is made up of a comparatively large number of photons and an appearance of continuity is given. At the other end of the spectrum, for example in the case of cosmic rays, the frequency is so high that the particle nature predominates.

1.2. *Emission and Absorption*

In 1817 Fraunhofer had observed dark lines in the spectrum of the sun. These were accounted for in 1861 by Bunsen and Kirchoff who proposed that light from the inner, hotter region of the sun consisted of waves having a continuous range of frequencies and which on passing through the cooler outer atmosphere underwent selective absorption. The dark lines superimposed on the continuous spectrum were thus characteristic of the outer atmosphere of the sun. Some of these lines could not be identified when compared with known spectra of pure gases and were therefore attributed to a new gas which was christened helium by Sir Norman Lockyer in 1868. It was not until 1894 that Sir William Ramsay discovered helium on earth.

The discovery of helium is a good example of the processes of emission and absorption and marked a period when a large amount of data was gathered on spectral lines but with no explanation on how different spectral series and their wavelengths came about.

By 1911 Rutherford had postulated that atoms could be described in terms of a central positively charged nucleus surrounded by a cloud of negatively charged electrons which orbited around the nucleus. In classical theory it was difficult to explain why the negatively charged electrons should not be attracted by the opposite charge on the nucleus and collapse onto it.

Niels Bohr in 1913 considered the simplest possible atom, namely that of hydrogen, which consists of a nucleus and a single electron. He suggested that the orbits of the electron were limited to a range of discrete sizes—no others being possible. The energy of the atom in turn took on certain discrete values according as to which orbit the electron occupied. In addition, when the atom reduced its energy by the electron moving to a different permitted orbit a photon was emitted whose energy ΔE was equal to the difference in the energy of the atom before and after the transition. The photon could be described in terms of a wave whose frequency v was given by Einstein's relation $\Delta E = hv$. Hence a collection of a large number of hydrogen atoms could only emit a spectrum of discrete lines whose frequencies were a result of all possible combinations of initial and final energy states of the single atom. In this way Bohr was able to explain quantitatively the emission spectra of hydrogen and similar simple atoms.

More complicated atoms involve more difficult calculations, but in general the simple picture of Bohr is still useful and the same principles hold : any atom can be described in terms of a set of possible energy levels, each level corresponding to a particular electron configuration.

For the sake of simplicity we shall consider an atom having only two possible energy states, an upper state E_2 and a lower state E_1 as shown in fig. 1.3.

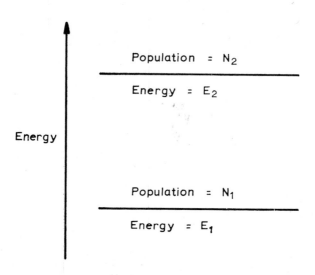

Population $=$ N$_2$

Energy $=$ E$_2$

Energy

Population $=$ N$_1$

Energy $=$ E$_1$

Fig. 1.3. Two-level energy system.

If the atom is in the upper state and makes a transition to the lower state then energy can be emitted in the form of radiation of frequency given by equation (1.3).

$$\nu = \frac{E_2 - E_1}{h} \qquad (1.3)$$

On the other hand, if the atom is initially in the lower energy state E_1 and makes a transition to the higher state E_2 then energy, and hence radiation of frequency given by the above equation, must be absorbed.

In 1917, in the course of calculations on the equilibrium of a gas, Einstein[3] discovered that there must be two possible types of emission. As we have seen emission involves the changing of the atom from a higher energy state to a lower state. We have not said anything about how this process comes about. In fact Einstein showed that emission can occur in two ways:

(i) by the atom changing to the lower state at random. This is called spontaneous emission and is contrasted with absorption in fig. 1.4.

8

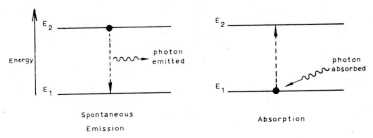

Fig. 1.4. Spontaneous emission and absorption.

(ii) by a photon having an energy equal to the energy difference
between the two levels interacting with the atom in the upper
state and causing it to change to the lower state with the creation
of a second photon.   This process can be thought of as the converse of absorption and is known as stimulated emission.   See
fig. 1.5.

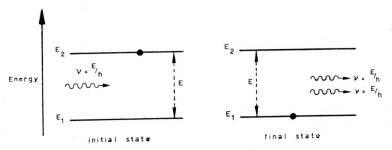

Fig. 1.5. Stimulated emission.

Now there are two very important points about stimulated emission
upon which the properties of laser light depend.   First the photon
produced by stimulated emission is of almost equal energy to that which
caused stimulated emission and hence the light waves associated with
them must be of nearly the same frequency.   Secondly the light
waves associated with the two photons are in phase—they are said to be
coherent.   In the case of spontaneous emission the random creation of
photons results in waves of random phase and the light is said to be
incoherent.   We shall consider these aspects further in Chapter 2 but
before continuing with our chronological account of the laser we must
consider how a collection of atoms distribute themselves among the
various possible energy states.

1.3. *Population Inversion*

Consider two two-level energy systems representing an atom in an
upper and lower state.   Suppose that a photon of energy equal to the

9

energy difference between the two levels approaches the two atoms: then which event is more likely to take place, absorption or stimulated emission? Einstein showed that, under normal circumstances, both processes are equally probable. It is thus apparent that in a system containing a very large number of atoms (or molecules) the dominant process will depend on the relative number of atoms in the upper and lower states. A larger population (i.e. number of atoms) in the upper level will result in stimulated emission dominating, while if there are more atoms in the lower level there will be more absorption than stimulated emission.

Under conditions of thermal equilibrium, the population of a number of energy levels obey what is known as a Boltzmann distribution, which means that for any two levels of energy E_1 and E_2 and population N_1 and N_2 then

$$\frac{N_2}{N_1} = \exp\left[-(E_2 - E_1)/kT\right] \tag{1.4}$$

where k is Boltzmann's constant, T is the absolute temperature and it is assumed that E_2 is greater than E_1.

If the energy difference between the two levels is kT, which is about 0·025 eV at room temperature, the population of the upper level will be i/e, or about 0·37, of the lower level. For a transition to the ground state to give rise to a photon of visible light the energy gap between the excited state and the ground state must be about 1·25 eV. Therefore under normal equilibrium conditions and at room temperature, the population of such an excited state will be almost negligible. Consequently any photons of visible light are very much more likely to be absorbed rather than to cause a stimulated emission.

For stimulated emission to dominate it is necessary to increase the population of the upper energy level so that it is greater than that of the lower—a situation known as a population inversion. It should be noted that equation 1.4 shows that even if the temperature is infinitely high, the populations will only be equal and absorption will just equal stimulated emission. For stimulated emission to exceed absorption, the temperature T in equation 1.4 would have to be negative, for this reason a state of population inversion is sometimes misleadingly referred to as a negative temperature. Population inversion can be achieved at ordinary temperatures but only under non-equilibrium conditions to which Boltzmann's Law does not apply.

1.4. The Ammonia Maser

The idea of amplification by means of stimulated emission occurred to several people after the Second World War including Webber[4], Fabrikant, Basov and Prokharov[5] and Townes. Charles Townes was, however, the first to build a device on this principle.

During the Second World War considerable effort was devoted to extending the wavelengths used in radar sets into the centimetre and millimetre regions. Townes had been working on 1·25 cm waves during the war and had observed that strong absorption took place in ammonia vapour, NH_3. Prompted by the work of Purcell and Pound[6] in 1951, who had observed some 'negative temperature' effects, it occurred to Townes that the ammonia molecule might itself act as a microwave source.

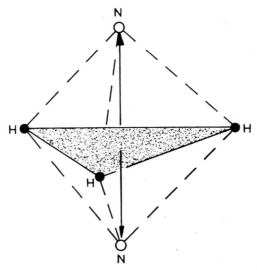

Fig. 1.6.   The ammonia molecule which consists of a nitrogen atom vibrating at right angles to the plane of the three hydrogen atoms.

An ammonia molecule can take on two energies according as to the position of the nitrogen atom with respect to the hydrogen atoms as shown in fig. 1.6. Townes argued that if, in a tank containing ammonia, sufficient atoms could be induced to exist in the higher state then a photon of energy equal to the difference between the two levels could trigger off a chain reaction in which stimulated emission dominated over absorption and built up an amplified wave by successive stimulated emissions. In the case of ammonia the energy gap involved would give a wave of frequency 23,870 MHz or a wavelength of about 1·25 cm. The amplification process is indicated in fig. 1.7.

Late in 1953 Gordon, Zeiger and Townes[7,8] at Columbia University constructed a device using this principle and christened it the maser, an acronym for *m*icrowave *a*mplification by the *s*timulated *e*mission of *r*adiation.

A diagram showing the essential components of this device is shown in fig. 1.8.

11

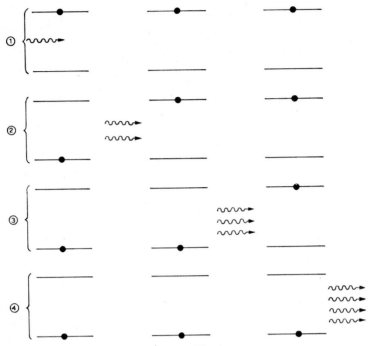

Fig. 1.7. The amplification process.

A directed supply of ammonia molecules is obtained by connecting a series of fine tubes to a tank containing ammonia gas at a pressure of a few torr. The beam of molecules is then passed through an electrostatic focusser which has the effect of converging those molecules in the high energy state and diverging those in the low energy state. The focusser itself consists of a cage of four rods 55 cm long and 1 cm apart which act as electrodes. Two opposite electrodes are earthed while the

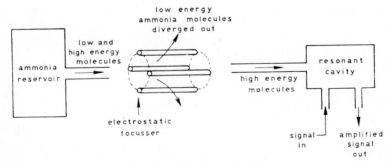

Fig. 1.8. The ammonia maser.

other two are kept at a potential of 15 kV. By this simple method of physically removing molecules in the lower state, a population inversion is achieved in the gas leaving the focussing system. The focussed beam of high energy molecules is then passed into a cavity designed to be resonant at a frequency corresponding to the energy gap between the upper and lower state of the ammonia molecule i.e. 23,870 MHz. Thus, if a microwave signal of this frequency is fed into the cavity, it can only stimulate downward transitions and is hence amplified. If enough high energy molecules are injected into the cavity then a spontaneous emission can start a self-sustaining chain reaction and the maser will then act as an oscillator.

The output power of the ammonia maser oscillator is very low, about 10^{-10} W, but of very high spectral purity (a line width of 5 Hz, or one part in 10^{10}, over a period of a minute is possible). As an amplifier, it is limited by a narrow band width of only a few kilohertz about the central frequency, and, of course, it is not possible to tune the device to work at different frequencies. For these reasons the ammonia maser is most useful as a frequency standard.

1.5. The Three-Level Maser

The search for a tunable maser of higher power and wider bandwidth led Bloembergen[9] to propose the idea of the three-level maser in 1956.

It has been explained that a necessary requirement for laser action to take place is the establishment of a population inversion between two levels of an atomic or molecular system. In the case of the ammonia maser this is achieved by the physical removal of molecules in the lower energy state from the system. Generally speaking such a technique is not possible and a more subtle way round the problem must be found.

Bloembergen's scheme was to select a suitable three-level system in which, as a result of the Boltzmann distribution, the population of each energy level would decrease from the bottom level to the top level and then to ‘ pump ’ atoms from the bottom level to the top by supplying photons of the correct frequency. As absorption is as equally possible as stimulated emission, the system would stabilize with the populations of the bottom and top levels equal. However, under these conditions, if the relative spacing between the three levels is carefully chosen, then the population of the middle state can be made to exceed that of the lowest. Consequently a population inversion between the middle and bottom states exists and so laser action is possible.

Unlike photons of light, the energy of a microwave photon is relatively very small in comparison with the thermal energy and hence the three energy levels will be relatively very close together and consequently very close to the ground state. It follows, therefore, that at room temperature their respective populations will be almost equal. Figure 1.9 a shows this situation. To increase the difference in population the active medium is cooled to liquid helium temperatures (4°K) thus

13

ensuring a large difference in population between the levels, of equation 1.4.

Figure 1.9 *b* shows the situation after cooling and fig. 1.9 *c* shows the situation after both cooling *and* pumping.

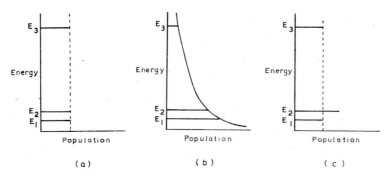

Fig. 1.9.   Energy level diagrams of the three-level maser.

It can be seen from the diagrams in fig. 1.9 that the pumping photons must have an energy $E_3 - E_1$ and that the maser output will consist of photons of energy $E_2 - E_1$.

It is apparent, therefore, that the three-level maser scheme overcomes the problem of any two level system where pumping can never result in a population inversion but only in equality of population.

In order to construct a three-level maser it is necessary to find a material which has three energy levels separated by small gaps equivalent to microwave frequencies. Bloembergen suggested that paramagnetic materials might be ideal for this purpose. These materials are crystals whose atoms or molecules are in effect permanent magnets and under the rules of quantum mechanics each atom or magnet can take up only discrete orientations with respect to an applied magnetic field. When the magnet points directly against the field the atom or molecule has the highest possible potential energy and when it points directly with the field the atom or molecule has the lowest possible potential energy.

The first device to operate on this principle was built by Scovil, Feher and Seidel[10] of the Bell Telephone Laboratories in 1956. This maser actually used the ion gadolinium in the form of gadolinium ethyl sulphate. The material was cooled to 1·2°K and pumped at a frequency of 17,500 MHz with an applied field of 2850 oersteds. Maser action was obtained at a frequency of 9060 MHz.

In the following year, 1957, McWhorter and Meyer[11] of the Lincoln Laboratory at the Massachusetts Institute of Technology used chromium ions in the form of doped potassium cobalticyanide and obtained oscillation and amplification. Figure 1.10 shows how the energy levels of the chromium ion split on application of the field and shows the pumping frequency of 9400 MHz and the maser transition at 2800 MHz. The device was operated at 1·25°K and a bandwidth of several hundred kilohertz was obtained.

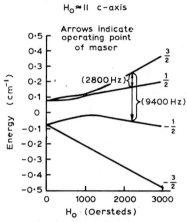

$H_0 \approx \parallel$ c-axis

Fig. 1.10. Variation of energy levels with applied magnetic field for chromium ions in potassium cobalticyanide (from reference 11).

Later in 1957 Makhov, Kikuchi, Lambe and Terhune[12] of the University of Michigan Engineering Research Institute obtained maser action in ruby. This material was subsequently used for masers built by many people including the GPO for the TV link on Goonhilly Down.

The masing ion in ruby is again chromium but in a host lattice of aluminium oxide, (Al_2O_3). Between 0·01 and 1% of chromium is added, the atoms of which take up positions as chromic ions (Cr^{3+}) within the crystal lattice.

1.6. *The Ruby Laser*

Naturally the advent of the maser brought about speculation as to whether such a device might be made to operate in the visible region of the spectrum. In 1958 Schawlow and Townes[13] put forward their proposals for extending the range of frequencies. They suggested that the resonant cavity might take the form of two parallel plane mirrors facing each other some distance apart. This form of resonator is a well known optical configuration and is called a Fabry-Perot interferometer. Schawlow and Townes suggested that the space in between the mirrors be filled with the active material of the maser or laser, as it was now to be called (the ' l ' standing for light).

In 1960 Maiman[14] of the Hughes Research Laboratory obtained pulsed laser action at 6943 Å in the red region of the spectrum using a ruby crystal as the active material. The Fabry-Perot resonator was made by simply polishing the ends of the ruby flat and parallel to a minute of

15

arc and then aluminising both ends. One end was made almost totally reflecting and the other about 10% transmitting in order to obtain some output from the device. The essential components of the ruby laser are shown in fig. 1.11.

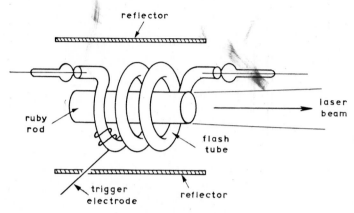

Fig. 1.11. The ruby laser.

The ruby laser of Maiman is an example of a three-level system (see fig. 1.12) operating on a somewhat different basis to the ruby maser. It should be noted the energy levels employed in the ruby laser are quite different from those used in the ruby maser. In the case of the laser they are also much further apart.

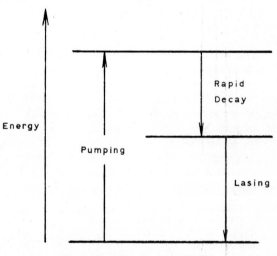

Fig. 1.12. The three-level laser.

16

A simple account of the mechanism of the ruby laser is now given ; a more complete account will be found in chapter 6.

Chromium ions in the ground state are excited to an upper state by an intense flash of white light. The upper state actually consists of a large number of levels forming a band. This makes the pumping much more efficient than pumping into a single level because more of the pumping radiation is utilised. Unlike the ruby maser situation, the excited atoms then drop back from the band of upper states to a middle state.

The transition down to the middle state is accompanied, not by the emission of a photon, but by the direct transfer of energy to the surrounding crystal lattice, which has the effect of heating up the ruby rod. This latter process is known as a non-radiative transition. Once an atom reaches the middle state it spends an unusually long time there before dropping down, by spontaneous emission, to the ground state. States such as the middle state are said to be metastable and it is because of this characteristic that the population of the middle state builds up while that of the ground state is depleted, i.e. a population inversion is achieved.

In a three level laser the method of obtaining population inversion between the middle and ground state is somewhat inefficient. This occurs because as the middle state is effectively empty at the start of pumping (as a result of the Boltzmann distribution and the large energy gap between the ground and middle states), at least half the population of the ground state (i.e. half the total number of atoms) must be pumped into the middle level before population inversion is achieved. In addition, very little of the electrical energy which is supplied to the flash lamp ends up as pumping photons and carefully designed reflectors round the ruby rod are essential. The intense pumping flash is necessarily brief and care must be taken to prevent the ruby rod from overheating. Consequently pulsed operation is always used although in 1962 Nelson and Boyle[15] of the Bell Telephone Laboratories demonstrated that continuous operation was possible if the device was cooled in liquid nitrogen.

1.7. *The Helium-Neon Laser*

The first continuously operating (c.w.–continuous wave) laser was constructed in 1960 by Javan, Bennett and Herriott[16] of the Bell Telephone Laboratories.

The laser action took place between two excited levels of neon, the function of the helium being to excite the neon atoms to a higher level—this process will be explained in detail in chapter 7. The source of excitation was a radio frequency field of 27 MHz. The helium-neon laser is essentially a four level system (see fig. 1.13).

Atoms in the ground state E_0 are excited to the highest level E_3 from which they descend non-radiactively, to a metastable state E_2. Providing level E_1 is sufficiently high above the ground state then it will be

17

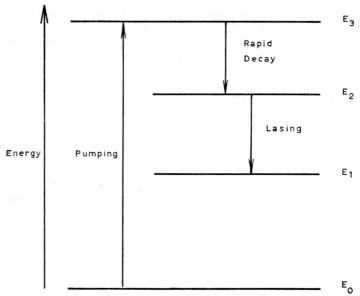

Fig. 1.13. The four-level laser.

effectively empty and so a comparatively small population in E_2 is needed to ensure a population inversion between E_2 and E_1. Laser action can therefore take place between these levels. For this reason a four-level laser is inherently much more efficient than a three-level laser and so continuous operation is much easier to obtain.

Javan's laser consisted of a quartz tube 100 cm long and 1·5 cm internal diameter and filled with a mixture of helium at 1 torr pressure and neon at 0·1 torr pressure, the length of the discharge in the tube

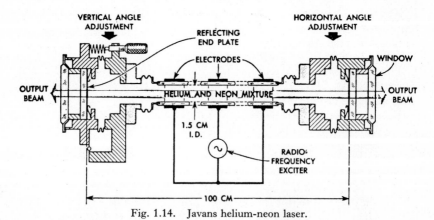

Fig. 1.14. Javans helium-neon laser.

18

being 80 cm. The general arrangement of the components is shown in fig. 1.14.

The ends of the tube were terminated with a pair of flat parallel mirrors each being 98·9% reflecting. Oscillation was obtained at five wavelengths in the infrared, the strongest output being at 1·1523 μm with a power of 15 mW when 50 W of r.f. power were applied to the external electrodes. Later on the now familiar 6328 Å output was also obtained.

By 1962 many research teams had started projects on lasers with the subsequent result that scores of different laser materials in the form of liquids, solids and gases have been found with literally thousands of lasing lines becoming available. Nevertheless, of all these, only a handful are to be found either in common use or commercially available.

CHAPTER 2

coherence

2.1. *Conditions for Coherence*

THE beam of light emitted by a laser can have the property of being almost completely coherent. Conventional sources of light, however, such as a fluorescent tube, a tungsten filament electric bulb or the sun are said to be incoherent. It will be explained that it is possible to take such an incoherent source and make it almost coherent but the light intensity obtained is so small as to be of very little use and for all practical purposes only the laser can give a powerful coherent beam of light. It is this previously unobtainable property that makes the laser such an important discovery in modern physics.

The light produced by a laser can be thought of as a wave oscillating some 10^{14} times a second and of wavelength about $1/100$ mm : for such a wave to be coherent two conditions must be fulfilled—first it must be of very nearly a single frequency, that is the spread in frequency or line-width must be small. If this condition holds then the light is said to have high temporal coherence. Secondly, the wavefront must have a shape which remains constant in time; if this condition holds, the light is said to be spatially coherent. A wavefront is defined to be the surface formed by points of equal phase. In the case of a point source of light, a wave is produced such that at any fixed distance from the source the phase is the same—thus a point source of light emits a spherical wavefront. Similarly a perfectly collimated beam of light has a flat wavefront.

A perfectly coherent source of light must be completely temporally and spatially coherent.

2.2. *Time Coherence*

In order to understand what is meant by time coherence and why it is associated with a narrow linewidth it is necessary to consider further how a photon of light is formed, the mechanism of which was outlined in the last chapter. Suppose an electron changes its orbit around an atom to a new one of lower energy, then a photon having a discrete energy E is emitted. The emission of a photon (or the change in energy of the atom) can be said to take place in a finite time which is denoted by Δt and called the lifetime. For atoms in free space, i.e. isolated from external influences, Δt has a value of about 10^{-8}s and so a photon can be visualized as a wave of finite length with an amplitude which rises from a value of zero to a peak and then drops again to zero when the electron has completed its transition to the lower energy level a lifetime later.   The frequency of the emitted wave is given by

$$\nu = E/h \qquad (2.1)$$

Such a wave is known as a Gaussian wavetrain and its shape is shown in fig. 2.1.

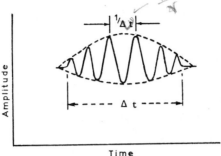

Fig. 2.1.   Gaussian wave train.

It should be understood that normally waves of many different frequencies are emitted by the atom after it is excited, with the result that the spectrum consists of a collection of single lines.   However, as far as this discussion is concerned, we are only considering just one line arising from one particular transition.

The light which is emitted from a gas discharge lamp will consist of a very large number of such wavetrains, each being radiated from those atoms in the gas which are undergoing transitions from higher to lower energy states.   The energy being continually lost from the lamp in this way is derived from the electrical energy supplied to it.   Now suppose for the sake of argument that it is possible with an imaginary instrument to watch some chosen point in the path of the wavetrains emitted by the lamp.   Suppose, also, that our instrument can detect changes in amplitude and phase over very small periods of time (better than say 10^{-14}s which is roughly the time taken for one cycle of the wave to pass the point of observation).   It is then possible, as each Gaussian wavetrain passes the observation point to watch the amplitude of the wave rise and fall.   If the frequency is 10^{14} Hz and Δt is 10^{-8}s then it is obvious that about one million undulations in amplitude will be observed.

As the atoms in a gas radiate in a random fashion, it is impossible to predict when the next wavetrain will arrive at the point of observation after the previous train has passed by.   Nevertheless, once the front end of a wavetrain has reached the observation point a prediction can be made about the amplitude and phase at some later time, assuming, of course, that the wavetrain is still passing the observation point.   It can therefore be seen that Δt is the longest time interval over which a prediction of phase and amplitude can be made.

21

It is this ability to predict amplitude and phase which is the essence of coherence. The light is said to be coherent for the time Δt and the longer this time the greater the coherence. The lifetime Δt is therefore also known as the coherence time and this type of coherence is referred to as time coherence.

The actual length of the wavetrain, $\mathscr{L}$, can be obtained by multiplying the number of cycles in the wavetrain by the wavelength, i.e.

$$\mathscr{L} = \frac{\Delta t}{1/v} \cdot \lambda \tag{2.2}$$

which is of course equivalent to multiplying the coherence time by the velocity of light,

$$\mathscr{L} = c\Delta t \tag{2.3}$$

At the beginning of this chapter it was stated that a narrow linewidth implies time coherence—this is readily apparent by applying a mathematical technique known as Fourier analysis to the Gaussian wavetrain. We have seen that the radiation emitted by an atom as a consequence of an electronic transition between two energy levels, consists of wavetrains of finite length and single frequency v. If, however, Fourier analysis is applied to one such wavetrain it can be shown[17] that it is equivalent to a number of infinitely long wavetrains of differing frequencies spread about a central frequency v. The spread, or linewidth, of these frequencies is measured at the half amplitude points and is denoted by Δv. The shape of the line is indicated in fig. 2.2. This is often a more convenient description of the finite wave train but it must be emphasized that both descriptions are equally valid.

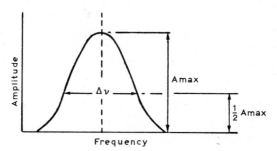

Fig. 2.2. Gaussian wave train.

It is useful to obtain a relation between Δv, the linewidth of the light source, and the coherence length, $\mathscr{L}$. To do this we make use of the expression[17] which sums up the above description of the Fourier analysis:

$$\Delta v = \frac{1}{\Delta t} \tag{2.4}$$

From this relation it can be seen that a source of light which generates long wavetrains having large values of Δt and hence great time coherence, has a very small linewidth. Thus long coherence time implies narrow linewidth.

Time coherence can therefore be expressed in three ways:

(a) the coherence time Δt, which is the time taken for the wavetrain to pass the point of observation,

or

(b) the linewidth $\Delta \nu$ (or $\Delta \lambda$)

or

(c) the length of the wavetrain or coherence length, $\mathscr{L}$.

The relation between coherence length and linewidth is derived as follows:
we have from equation 2.3

$$\mathscr{L} = c\Delta t \tag{2.5}$$

hence from equation 2.4

$$\mathscr{L} = c/\Delta \nu \tag{2.6}$$

but

$$c = \nu \lambda \tag{2.7}$$

differentiating equation 2.7 we have

$$\Delta \nu = \frac{c}{\lambda^2} \Delta \lambda \tag{2.8}$$

(ignoring the negative sign) and substituting equation 2.8 in equation 2.6 we finally obtain

$$\mathscr{L} = \frac{\lambda^2}{\Delta \lambda} \tag{2.9}$$

We can now calculate the coherence length of a wavetrain resulting from a typical free space atomic energy level transition having a lifetime of 10^{-8}s. We have

$$\Delta t \sim 10^{-8}\text{s} \tag{2.10}$$

$$\therefore \Delta \nu \sim 10^8 \text{ Hz from equation 2.4} \tag{2.11}$$

If we assume green light for which

$$\lambda = 0 \cdot 5 \times 10^{-4} \text{ cm} \tag{2.12}$$

then from equation 2.8

$$\Delta \lambda = 0 \cdot 001 \text{ Å} \tag{2.13}$$

from which finally

$$\mathscr{L} \sim 3 \text{ metres} \tag{2.14}$$

23

This result is obtained for the ideal case of a single atom. When a collection of atoms are in close proximity the natural linewidth, as the linewidth corresponding to the former situation is called, is broadened out. This broadening of the linewidth has two principal causes; the presence of other radiating atoms (pressure broadening) and the motion of other radiating atoms (Doppler broadening). Despite these effects a low pressure cadmium lamp can give a line whose width is only 0·01 Å with a resulting coherence length of 30 cm. Most normal sources of light, however, consist of a large number of broader lines each having a coherence length which is very small indeed—perhaps a fraction of a millimetre.

For solid materials the lines may be broadened out into a continuum as the atoms are so close that electric and magnetic fields associated with the atoms distort the energy levels to a considerable and widely varying degree. This has an important bearing on the output of solid state lasers which will be referred to later.

The coherence lengths available from lasers will be considered in detail later in the next chapter. Suffice it to say that coherence lengths of many hundreds of kilometres are possible although generally speaking most commercially available lasers have outputs whose coherence lengths range from a few centimetres to tens of metres.

2.3. *Space Coherence*

For a source to be coherent it must have time and space coherence. We must now consider the subject of space coherence in greater detail.

Time coherence implies the possibility of predicting phase and amplitude after some given time interval between initial and final observations. If this process of prediction can be repeated at some later time, then time coherence of magnitude equal to the intervals between predictions is said to exist. In the case of spatial coherence, however, we are not concerned with different observations at different points in time along the wavetrain but in different points, in space, on the wavefront. A wave is said to be spatially coherent if there is a constant phase difference between any two chosen points on the wavefront. The term ' constant ' implies a time long enough to perform some operation on the wavefront such as observation with the eye or photography. Well designed lasers can maintain spatial coherence almost indefinitely in contrast to time coherence which usually can be held only for a fraction of a second.

Conventional light sources are always incoherent as a result of the random uncoordinated emission of photons by the atoms of the source. In the case of lasers, the situation is quite different. This is because the photons are created as a result of chain reactions of stimulated emissions in which all the photons produced represent waves *in phase*.

Time and space coherence then can be summarized as follows: when we speak of time coherence we are saying that the relative phases

24

between two points *in time* must remain constant over some long time interval: space coherence, on the other hand, involves the relative phases between two points *in space* remaining constant again over some long time interval. In each case, the longer the time interval the greater the coherence.

In order to clarify these two concepts in practical terms it will be helpful to consider two classical experiments, one performed by Michelson and the other by Young.

2.4. *The Michelson Interferometer*

Figure 2.3 shows the optical arrangement called the Michelson Interferometer. A wavefront is produced by the source S and is divided into two parts by the partially silvered mirror B to provide two wavefronts of identical shape and intensity. The two mirrors M_1 and M_2 serve to recombine the two wavefronts at the observation point O which is usually a telescope.

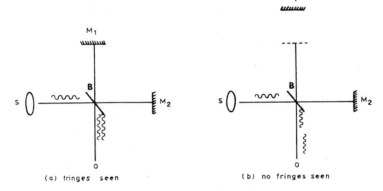

Fig. 2.3.   The Michelson Interferometer.

Consider a wavetrain emitted from the extended source S. If the distance between the beam splitter and the two mirrors is the same (the beam splitter being assumed infinitely thin) then the two wavetrains produced by the beam splitter will overlap exactly as they travel from the beam splitter to the observation point O. As a result fringes will be seen on looking through the telescope (see fig. 2.3 *a*). If, however, a time delay is introduced into one of the wavetrains, exact overlap will not occur at the telescope. The result of this is that the fringes seen will be of reduced contrast because that part of the wavetrain over which overlap has not occurred will have the effect of adding a bright background to the fringe system.

If the mirror M_1 is moved to M_1' (see fig. 2.3 *b*) so that the path traversed by one wavetrain is longer (or shorter) than that traversed by the other by at least its own length, then no interference fringes will be seen. Uniform illumination will be observed through the telescope.

25

If either mirror is moved perpendicular to its plane so that fringes are just not seen, then the distance through which the mirror has been moved is half the length of the wavetrain, in this way the length of the wavetrain, i.e. the coherence length, is easily measured. From this can be deduced the coherence time, this being the time taken for the wavetrain to traverse the greatest possible extra path length consistent with fringes still being visible.

An extended source of light is used in the Michelson Interferometer so that fringes can be seen over a reasonable angle on looking through the telescope. Spatial coherence of the source is *not* necessary as the *relative* difference in phase between the recombined wavefronts is constant no matter how rapidly their absolute values may fluctuate.

2.5. *Young's Two-Slit Experiment*

The two-slit experiment described in chapter one as being responsible for the acceptance of a wave theory of light is a good example of the necessity for spatial coherence for interference to occur. In Young's experiment, shown in fig. 2.4, two sources of light are emitting waves which combine to form fringes on a screen.

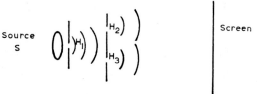

Fig. 2.4. Young's two-slit experiment.

The spacing of the fringes formed by two slits can be easily calculated with reference to fig. 2.5

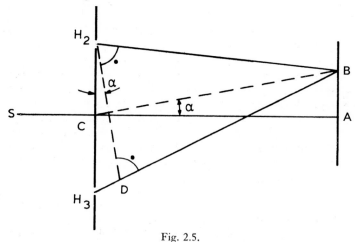

Fig. 2.5.

26

H_2 and H_3 are two slits which act as sources assumed to be in phase.   If the slits are symmetrically positioned about the line SA (the optic axis) then a bright fringe will be observed on the screen at A.   Suppose that a new position B is chosen such that AB subtends an angle α at the point C, then a discrepency in path difference between rays travelling from H_1 to B and H_2 to B is introduced.   If AB is small compared with AC then

$$\alpha = \frac{AB}{AC} \qquad (2.15)$$

$$\text{but} \qquad \alpha = \frac{H_3 D}{H_2 H_3} \qquad (2.16)$$

If B is chosen to be at the position of the first bright fringe away from that at the centre, then

$$H_3 D = \lambda \qquad (2.17)$$

and so substituting equation 2.17 into equation 2.16 and using equation 2.15 we obtain

$$\frac{\lambda}{H_2 H_3} = \frac{AB}{AC}$$

let the slit separation $\qquad H_2 H_3 = d$

and let $\qquad\qquad\qquad AC = f$

then the fringe spacing is given by

$$AB = \frac{\lambda f}{d} \qquad (2.18)$$

if $\lambda = 6328 \text{ Å}\, f = 1$ m and $d = 1$ mm, then the fringe separation would be approximately 0·6 mm.

With a conventional, extended, incoherent source it is necessary to place a small pinhole at H_1 which acts as a coherent point source of light to illuminate the two slits at H_2 and H_3.   It is assumed that a coloured filter is placed in front of the source so that essentially monochromatic light is used. H_2 and H_3 act as two coherent sources and a set of approximately straight line fringes appear on the screen.   If no filter were used and the source was white light then coloured fringes would be obtained which would overlap and tend to blur out the pattern.

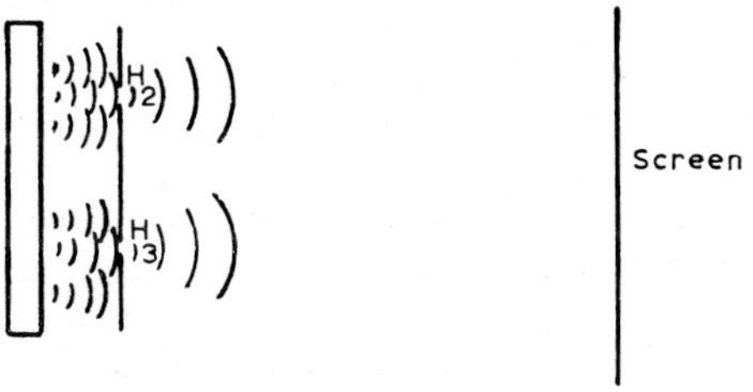

Fig. 2.6.   Young's two-slit experiment using a fluorescent tube as the source.

Suppose that two slits H_2 and H_3 are illuminated by an incoherent source such as a fluorescent tube as shown in fig. 2.6. Even if a filter was placed in front of the tube to provide a monochromatic source no fringes would be seen. If it were possible to take a snapshot of the screen in 10^{-15}s then a fringe system would be recorded, but another snapshot of similar duration taken at a later time would show fringes in different positions due to the change in relative phase between H_2 and H_3. Thus over a 'long' period of time the very rapidly changing fringes average out to produce uniform illumination as far as the human eye is concerned.

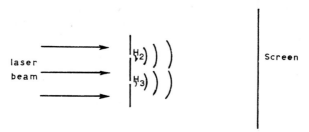

Fig. 2.7. Young's two-slit experiment using a laser beam as the source.

Figure 2.7 shows the tube replaced by a laser which is emitting a plane wavefront. In this case fringes are seen because the holes H_2 and H_3 are illuminated by a spatially coherent wavefront. That is to say H_2 and H_3 act as sources in phase.

In fig. 2.7 a plane wavefront is shown illuminating the slits. If a ground glass diffusing screen were placed between the laser and slits would fringes still be seen ? The answer is yes because the phase at the two slits does not have to be the same, only of constant difference. This brings out a point over which there is sometimes confusion. Spatial coherence does not necessarily mean that the wavefront is flat, only that the shape of the wavefront does not vary in time. A flat wavefront is a special case of spatial coherence where the phase is constant over the wavefront and is referred to as *uniphase*. If the ground glass screen placed in front of the laser in fig. 2.7 were rapidly rotated then the light would become spatially incoherent and no fringes would be seen.

2.6. Coherent Waves from Incoherent Sources

Figure 2.8 shows how a uniphase temporally coherent wave can be obtained from an incoherent source by taking a low pressure lamp and placing a small pinhole in front of it. The light is filtered so as to select only one wavelength. If a lens is then placed such that its focus is at the pinhole then a plane (uniphase) wavefront is produced. However, it must be remembered that such a wavefront would be of extremely small intensity—many orders less than that produced by a laser.

28

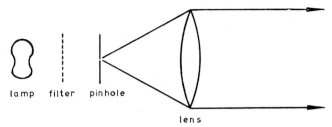

l a m p   f i l t e r   p i n h o l e

l e n s

Fig. 2.8.   Production of coherent light from a conventional source.

The size of the hole used to spatially filter the incoherent light is of critical importance.   If the ' hole ' is actually a slit, the maximum width of the slit is obtained as follows :

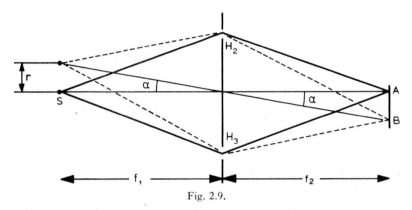

Fig. 2.9.

With reference to fig. 2.9, H_2 and H_3 are again two slits distance d apart symmetrically placed with respect to SA and a bright fringe is formed at A.   Suppose the point S is now moved perpendicular to SA (i.e. the slit is increased in width), so moving the central bright fringe on the screen away from A to a new position B.   If the movement r is sufficient, the bright fringe will be superimposed on the original position of the first dark fringe.   Let r be of such magnitude as to displace the first bright fringe through half a fringe spacing.   Consequently if the width of the slit source is increased until it is just equal to $2r$, the fringes will disappear completely and the source is then said to be just incoherent.   The fringe spacing is given by equation 2.18, hence for the source to be just incoherent :

$$AB = \tfrac{1}{2}\frac{\lambda f_2}{d} \tag{2.19}$$

but

$$\alpha = \frac{AB}{f_2} = \frac{r}{f_1} \tag{2.20}$$

29

$$\therefore \qquad \frac{rf_2}{f_1} = \frac{1}{2}\frac{\lambda f_2}{d} \qquad (2.21)$$

$$\therefore \qquad r = 0\cdot5\frac{\lambda f_1}{d} \qquad (2.22)$$

i.e. the width of the slit must be less than $\lambda f_1/d$ to act as a coherent source. For a circular source of radius r, it can be shown that for the source to be just incoherent[1]

$$r = 0\cdot61\frac{\lambda f_1}{d} \qquad (2.23)$$

The area within this diameter is sometimes referred to as the *coherence area*.

It may be shown that[18] an *almost* completely coherent source of circular shape is that whose radius r is given by

$$r = 0\cdot16\frac{\lambda f_1}{d} \qquad (2.24)$$

Thus, in a Young's two-slit experiment, if for example light of wavelength 5000 Å is used and the slits are 1 mm apart and 10 cm from the screen, then the maximum diameter of a circular source is, from equation 2.24, 16 μm.

Similarly, if an f 2·8 lens is used to produce a plane coherent wave-front by collimating light of 5000 Å wavelength from a pinhole placed in front of an extended source, the pinhole must not be more than 0·45 μm in diameter.

An interesting and instructive application of incoherent sources appearing coherent, if they are sufficiently small, is in the measurement of the angular diameter of stars. Suppose a star is imaged by means of a telescope with two Young's slits placed over the objective as shown in fig. 2.10.

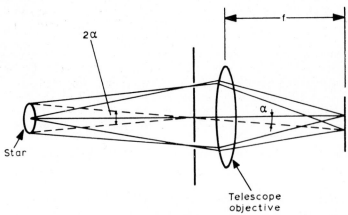

Fig. 2.10.

Then the image of the star will be a diffraction pattern of finite size crossed by a set of straight line fringes. If the slits are gradually moved apart, the fringes will eventually disappear. If the separation of the slits in this position is h, then from equation 2.22

$$\alpha = \tfrac{1}{2} \cdot \frac{\lambda f / h}{f} \qquad\qquad (2.25)$$

i.e.
$$\alpha = \frac{\lambda}{2h} \qquad\qquad (2.26)$$

∴ the angular diameter of the star is given by

$$2\alpha = \frac{\lambda}{h} \qquad\qquad (2.27)$$

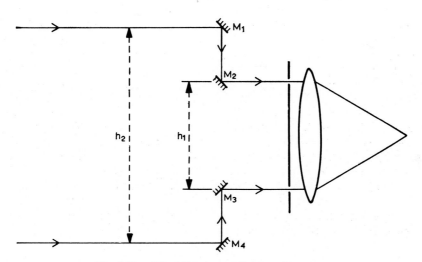

Fig. 2.11.   The Michelson stellar interferometer.

In practice the arrangement used is depicted in fig. 2.11 so as to avoid the necessity for very large objectives. The light from the star is reflected via four plane mirrors M_1, M_2, M_3, and M_4, thus effectively increasing the spacing of the slits from h_1 to h_2 and so increasing the angular resolution. Such a system is called a Michelson stellar interferometer after its inventor. More recently an electronic equivalent of the Michelson stellar inferometer, known as a Hanbury–Brown and Twiss stellar interferometer[19], has been developed.

CHAPTER 3

some laser theory

IN the first chapter of this book the principles underlying laser operation were outlined in a general fashion. We shall now discuss some important quantitative relations which indicate the relative importance of the various parameters involved.

3.1. *Black-Body Radiation*

All bodies radiate energy. This radiation arises from the motion of the atoms which make up the body, hotter bodies radiating more energy than cooler ones. In 1879 Stefan, as a result of his measurements on radiation, proposed a law that related the total energy radiated from a body to its temperature. Stefan's law can be expressed by means of the following equation :

$$R = \sigma e T^4 \qquad (3.1)$$

where R is the total thermal radiation per unit surface area of the body, σ is a constant called Stefan's constant, T is the absolute temperature and e is a number between 0 and 1 which depends on the nature of the emitting surface—this number is known as the emissivity.

By allowing a beam of radiation to fall onto various bodies and studying the radiation absorbed, Kirchhoff, in 1895, showed that the ratio of the energy incident on the body to that absorbed by the body, a, was related to the emissivity quite simply by the equation :

$$e = a \qquad (3.2)$$

where a is called the absorbtivity.

The radiation from the sun forms a continuous spectrum with dark lines superimposed on it ; as explained in the first chapter, these occur because of absorption by the cooler outer layers of the sun. It is of considerable theoretical interest to study radiation from a body which has not undergone any absorption at any frequencies. In other words the body is a perfect emitter of radiation ($e = 1$) at all frequencies. Such an ideal body is called a black-body, and the radiation it emits is known as black-body radiation. In the laboratory this can be simulated by forming a hollow cavity within a solid body and drilling a small hole to join the cavity to the exterior. If the body is kept at a constant temperature then the radiation emitted from the hole will be a very good approximation to black-body radiation. By supplying energy to the body to make up for that lost from the cavity, a state of thermal equilibrium is established. The total radiation emitted by a black-body is, from Stefan's law :

$$R = \sigma T^4 \qquad (3.3)$$

It is of much more interest, however, to find out what amount of radiation can be expected from a black-body at any particular frequency. This can be done experimentally and graphs of the form shown in fig. 1.2 have been obtained, these being plots of energy against frequency for various different temperatures.

The area under each graph is proportional to the total energy emitted by the black-body at that particular temperature. As the temperature increases, the total energy increases in agreement with Stefan's law and the character of the radiation becomes shorter in wavelength. When a solid body, such as a piece of iron, is heated, at first it does not change in colour because nearly all the radiation emitted is in the far infrared region of the spectrum to which the eye is not sensitive. As the temperature rises, the body becomes red hot and ultimately white hot. Under certain conditions it is possible for a body to be so hot that it appears blue ; an example of this is the very hot type of star known as a ' blue giant ' because of its distinctly blue appearance.

For some years before Planck's theory, many physicists, including Rayleigh and Jeans, had tried to work out a mathematical expression for the shapes of the curves shown in fig. 1.2 but without success. The advent of the quantum theory at last enabled a successful equation to be established. Without following the analysis in detail we shall indicate briefly the steps involved. First, imagine a cavity containing black-body radiation in thermal equilibrium. The radiation will take the form of standing waves within the cavity ; a particular wavelength, or mode of oscillation, is possible if its nodes, or points where the amplitude of the wave is always zero, coincide with the walls of the cavity. Three different modes of oscillation are indicated in fig. 3.1.

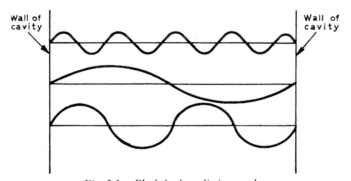

Fig. 3.1.   Black-body radiation modes.

The simplest mode consists of two nodes at the walls of the cavity. As the wavelength of the radiation corresponds to the distance between

33

alternate nodes it can be deduced that a condition for a mode of oscillation to exist is:

$$\tfrac{1}{2} n\lambda = L \qquad (3.4)$$

where λ is the wavelength, n is a positive integer and L is the length of the cavity.

The energy per frequency interval for a cavity of unit volume, which is what we wish to find, is obtained by multiplying the number of modes of oscillation per frequency interval by the average energy of a mode. In order to calculate the latter we must return to the Boltzmann distribution introduced in the first chapter, which tells us the relative population of different energy levels in a system in thermal equilibrium. The population N_1 of an energy level E_1 is given by :

$$N_1 = A \exp\left(- E_1/kT\right) \qquad (3.5)$$

where A is a constant. The population, N_2 of E_2 is similarly given by :

$$N_2 = A \exp\left(- E_2/kT\right) \qquad (3.6)$$

If therefore two energy levels E_1 and E_2 exist where E_2 is greater than E_1, then :

$$\frac{N_1}{N_2} = \frac{A \exp\left(- E_1/kT\right)}{A \exp\left(- E_2/kT\right)} \qquad (3.7)$$

from which it follows that :

$$\frac{N_1}{N_2} = \exp\left(\frac{E_2 - E_1}{kT}\right) \qquad (3.8)$$

In some instances an atomic energy level really consists of a number of distinct energy levels which happen to coincide in their energy values. The energy level is then said to be degenerate ; the symbol g indicates the number of energy states superimposed on one level and the level is said to have a degeneracy of g. Therefore equations 3.5 and 3.6 are written more correctly as

$$\frac{N_1}{g_1} = A \exp\left(-E_1/kT\right) \qquad (3.9)$$

and

$$\frac{N_2}{g_2} = A \exp\left(-E_2/kT\right) \qquad (3.10)$$

and therefore it follows that equation 3.7 is again written more correctly as :

$$\frac{N_1}{N_2} = \frac{g_1}{g_2} \exp\left(\frac{E_2 - E_1}{kT}\right) \qquad (3.11)$$

By using Planck's theory it is possible to calculate the average energy of each mode. According to Planck, the energy of a radiation field of frequency ν can take only the following values; zero, $h\nu$, $2h\nu$, $3h\nu$, etc., and from the Boltzmann distribution the number of oscillators having these energies is A, $A \exp\left(-\dfrac{h\nu}{kT}\right)$, $A \exp\left(-\dfrac{2h\nu}{kT}\right)$, etc. The average energy of a black-body radiation mode, ε, is hence simply obtained by dividing the total energy by the total number of oscillators i.e. :

$$\varepsilon = \frac{0.A + h\nu A \exp(-h\nu/kT) + 2h\nu\, A \exp(-2h\nu/kT) + \ldots}{A + A \exp(-h\nu/kT) + A \exp(-2h\nu/kT) + \ldots} \quad (3.12)$$

On simplifying :

$$\varepsilon = \frac{h\nu \exp(-h\nu/kT)[1 + 2 \exp(-h\nu/kT) + 3 \exp(-2h\nu/kT) + \ldots]}{1 + \exp(-h\nu/kT) + \exp(-2h\nu/kT) + \ldots}$$

$$(3.13)$$

The numerator is easily reduced by the binomial theorem and the denominator is a geometric progression and so equation 3.13 can be expressed in the much simplified form :

$$\varepsilon = \frac{h\nu}{\exp(h\nu/kT) - 1} \quad (3.14)$$

The number of modes per unit volume of the cavity per unit frequency interval is obtained by a straightforward procedure which can be found in physics textbooks, and is found to be :

$$\frac{8\pi \nu^2}{c^3} \quad (3.15)$$

and so the energy density, i.e. energy per unit volume u_ν of the cavity, per unit frequency interval, is given by :

$$u_\nu = \frac{8\pi h \nu^3}{c^3} \frac{1}{\exp(h\nu/kT) - 1} \quad (3.16)$$

If u_ν is plotted, as a graph against frequency ν, then curves identical to those shown in fig. 1.2 are obtained, a different curve being obtained for each temperature.

An interesting comparison can be made between the radiation emitted from a black-body and that from a laser. A ruby laser gives an output at 6943 Å in the red part of the spectrum and is capable of giving pulses of about 1 MW cm^{-2} with a line-width of about 0·1 Å. A black-body whose radiation peaks at 6943 Å would have to be at a temperature of 4174°K and in a region 0·1 Å wide about this peak the radiation flux is only about 16 mW cm^{-2} which indicates the enormous power available from pulsed lasers compared with that from a hot body.

Even when working continuously, a laser is still superior to a black-body. A helium-neon laser can give 10 mW cm^{-2} with a linewidth of 10^{-5} Å at 6328 Å. A black-body would have to be at a temperature of 4580°K to be most efficient at producing radiation at 6328 Å and would then only deliver 2·6 μW cm^{-2} in a corresponding spectral width.

3.2. The Einstein Coefficients

The lifetime of an atom is the average time it exists in an excited state before it makes a spontaneous transition to a lower energy state. As was stated in Chapter 1, an excited atom in free space has a lifetime of about 10^{-8}s. Another way of visualising this is to consider the reciprocal of the lifetime which is simply equivalent to the average number of spontaneous transitions from an excited state within a period of a second, and so in the case of the free state atom 10^8 atoms will make such transitions per second on average. This transition rate is synonymous with transition probability—the larger the transition rate the greater the probability of transition. The probability of a spontaneous emission occurring denoted by A, is called the Einstein A coefficient. It follows that :

$$A_{21} = \frac{1}{t_{21}} \tag{3.17}$$

the suffix 21 indicating a transition from an upper level labelled 2 to a lower level labelled 1, t_{21} is the lifetime and is equivalent to the coherence time Δt introduced in Chapter 2.

In a similar fashion two other Einstein coefficients can be defined for the cases of stimulated emission and absorption. They are termed B_{21} and B_{12}, again the order of the numbers indicates the initial and final states. The same letter B is used for both B coefficients as it is found that B_{21} and B_{12} are in fact identical. By using the black-body radiation formula (equation 3.16) we shall now give a proof of this and in addition derive a relation between the Einstein A and B coefficients.

In the two-level energy system shown in fig. 1.3 N_2 and N_1 are the populations of levels E_2 and E_1 respectively. The total number of spontaneous downward transitions from level 2 to level 1 each second is :

$$N_2 A_{21} \tag{3.18}$$

Photons of energy $h\nu$ arriving to interact with the system will either be absorbed or will stimulate emission. In each case the greater the number of photons present, i.e. the greater the radiation density, which is denoted by $\rho(\nu)$, the more likelihood of such transitions taking place. For absorption the total number of absorptions taking place per second will be given by :

$$N_1 B_{12} \rho(\nu) \tag{3.19}$$

36

and for stimulated emission the total number of stimulated transitions from the upper level to the lower level will be :

$$N_2 B_{21} \rho(\nu) \tag{3.20}$$

Suppose now that the system is in thermal equilibrium. This means that the total energy of the system must remain constant or, in other words, the number of photons absorbed per second must be equal to the total number emitted from the system by stimulated and spontaneous emission. It thus follows from equations 3.18, 3.19, and 3.20 that

$$N_1 B_{12} \rho(\nu) = N_2 A_{21} + N_2 B_{21} \rho(\nu) \tag{3.21}$$

from which :

$$\rho(\nu) = \frac{N_2 A_{21}}{N_1 B_{12} - N_2 B_{21}} \tag{3.22}$$

however, by the Boltzmann distribution which applies to systems in thermal equilibrium we have :

$$\frac{N_2}{N_1} = \exp\left(-h\nu/kT\right) \tag{3.23}$$

assuming that the degeneracy of each level is unity.
Substitution of equation 3.23 into equation 3.22 gives :

$$\rho(\nu) = \frac{A_{21}}{\exp\left(h\nu/kT\right)B_{12} - B_{21}} \tag{3.24}$$

Now an ideal two-level system such as we have been considering must result in a radiation density $\rho(\nu)$ which is identical to black-body radiation density. By comparing the above equation 3.24 with the black-body radiation equation 3.16 it follows, because the two expressions for $\rho(\nu)$ must be identical, that :

$$B_{12} = B_{21} = B \tag{3.25}$$

and

$$\frac{A_{21}}{B} = \frac{8\pi h \nu^3}{c^3} \tag{3.26}$$

It was, in fact, the necessity of establishing this identity that led Einstein to introduce the concept of stimulated emission in 1917.

Equation 3.25 shows that the probability of stimulation down is equal to the probability of absorption which seems intuitively sensible if absorption is thought of as 'stimulation up'. Equation 3.26 has some important implications ; the ratio R of the rate of spontaneous emission to the rate of stimulated emission under conditions of thermal equilibrium is given by :

$$R = \frac{A_{21}}{\rho(v)B} \qquad (3.27)$$

which on substitututing equation 3.24 into equation 3.27 becomes

$$R = \exp(hv/kT) - 1 \qquad (3.28)$$

If v is taken to correspond to the frequency of green light, i.e. 5×10^{14} Hz, then R is found to be equal to about e^{82} or roughly 10^{35} ! In other words the likelihood of a stimulated emission taking place is completely negligible compared to the probability of a spontaneous emission. If a frequency corresponding to a microwave transition is taken, e.g. 10^9 Hz, then R becomes about $0\cdot001$—a complete reverse in the probability situation. Radiowaves and microwaves arise almost entirely from stimulated transitions and so are always coherent. In all cases spontaneous emission manifests itself as undesirable noise within the system. It is also of interest to note that the rates of spontaneous and stimulated emission become equal i.e. $R = 1$ for a wavelength of about 60 μm in the far infrared region of the spectrum.

In this section we have assumed that thermal equilibrium exists and under such conditions the possibility of stimulated emission in the visible region of the spectrum is entirely negligible. By creating a population inversion, the thermal equilibrium is effectively destroyed and considerable stimulated emission of visible light becomes possible.

3.3 The Threshold Condition

From the point of view of efficiency it is clearly of importance to choose laser materials which require the minimum amount of energy to induce laser action. In practice this means finding out what population inversion must be achieved before all the the losses in the system are overcome. A laser consists of an amplifying medium, usually a gas or a solid, placed between two mirrors which form an optical resonator. The losses in this system are replenished by induced transitions occuring within the amplifying medium. The total loss is due to a number of different processes, the most important of which are :

(1) Transmission, absorption and scattering by the mirrors.
(2) Absorption within the amplifying medium due to other energy levels—no system is an ideal two-level one.
(3) Scattering by optical inhomogeneities within the amplifying medium—a particularly important source of loss in the case of solid-state lasers where it is impossible to provide a perfect crystal.

(4) Diffraction losses by the mirrors—this important aspect is discussed further in Chapter 4.

All these losses can be included in one parameter which can be expressed as the lifetime of a photon existing within the laser cavity. The amplifying medium is assumed to be neutral, i.e. the photon is assumed not to undergo any processes which are concerned with the two-energy levels between which laser action takes place. Such a lifetime is denoted by the symbol t_{photon}. The reciprocal of this will be the total rate of loss of photons from the laser per second as a result of the four processes outlined above.

Consider a two-level energy system. Denoting the stimulated transition rate for the time being by

$$W' = \rho(\nu)B \qquad (3.29)$$

Substituting from equation 3.26 it follows that :

$$\frac{W'}{A} = \rho(\nu)\frac{c^3}{8\pi h\nu^3} \qquad (3.30)$$

where the suffix 21 has been dropped from the term A_{21}.

It was mentioned previously that the laser linewidth is very narrow indeed—much narrower, in fact, than the linewidth of the atomic transition. The shape of the absorption curve is denoted by $g(\nu)$ and consideration will show that $g(\nu)$ can also be described as an emission curve or a frequency probability curve. We can define $g(\nu)d\nu$ as the probability that a given transition between energy levels will result in an emission, or an absorption, of a photon whose energy lies between $h\nu$ and $h(\nu + d\nu)$. The curve $g(\nu)$ is normalized so that the total area under it is always unity. The effect of all this is that a photon of frequency ν will not, with absolute certainty, stimulate another photon of frequency ν : it can only be stated that there will be a finite probability $g(\nu)d\nu$ that the stimulated photon will have a frequency which is between ν and $\nu + d\nu$. Similar arguments apply to spontaneous emission and it is therefore necessary to replace W' and A in equation 3.30 by $W'g(\nu)d\nu$ and $Ag(\nu)d\nu$ respectively. $W'g(\nu)d\nu$ is now the rate at which transitions take place resulting in photons of frequency lying between ν and $\nu + d\nu$ due to a radiation density $\rho(\nu)$. $Ag(\nu)d\nu$ is the rate of spontaneous emission into the frequency interval lying between ν and $\nu + d\nu$. Equation 3.30 may therefore be rewritten to give

$$W'g(\nu)d\nu = \frac{\rho(\nu)c^3}{8\pi h\nu^3}Ag(\nu)d\nu \qquad (3.31)$$

The radiation density $\rho(\nu)$ is simply related to the radiation flux or intensity $I(\nu)$ by :

$$I(\nu) = c\rho(\nu) \qquad (3.32)$$

where c is strictly speaking the velocity of light in the laser medium.

Using the fact that :

$$A = \frac{1}{t_{spont.}} \qquad (3.33)$$

where $t_{spont.}$ is the spontaneous emission lifetime, it follows that on integration of each side to give the total transition rate due to a monochromatic beam of radiation of frequency denoted by I_ν, equation 3.31 becomes

$$W(\nu) = \frac{c^2}{8\pi h \nu^3 t_{spont.}} g(\nu) I_\nu \qquad (3.34)$$

We are now in a position to write down the increase in intensity due to stimulated emission, but before doing so we must allow for any degeneracy in energy levels. Degeneracies of unity were assumed throughout the analysis of the Einstein coefficients. Suppose a degeneracy of g_1 is associated with the lower level E_1 and a degeneracy of g_2 with the upper level E_2 (this is standard notation and should not be confused with $g(\nu)$, the lineshape function). Bearing these degeneracies in mind, the net rate of change in energy due to stimulated transitions (both up and down) is obtained by multiplying the transition rate by the net population inversion and the photon energy $h\nu$ i.e. :

$$h\nu(N_2 - N_1 g_2/g_1) W(\nu) \qquad (3.35)$$

therefore the rate of increase in intensity may be expressed as :

$$\left(\frac{dI}{dt}\right)_{gain} = h\nu\left(N_2 - N_1\frac{g_2}{g_1}\right) W(\nu)c \qquad (3.36)$$

A gain or increase in intensity is assumed because a population inversion is also assumed i.e. :

$$N_2 \geqslant N_1\frac{g_2}{g_1} \qquad (3.37)$$

If this condition does not hold the laser medium will simply absorb which is what normally happens in any material without population inversion. The total loss rate is defined by means of the parameter t_{photon} discussed above :

$$\left(\frac{dI}{dt}\right)_{loss} = \frac{I\nu}{t_{photon}} \qquad (3.38)$$

combining equations 3.36 and 3.38 we obtain a condition which clearly must be satisfied for laser action to take place :

$$\left(\frac{dI}{dt}\right)_{gain} - \left(\frac{dI}{dt}\right)_{loss} \geqslant 0 \qquad (3.39)$$

40

or writing $v = v_0$ and substituting from equation 3.34

$$\left(N_2 - N_1 \frac{g_2}{g_1}\right) \geqslant \frac{8\pi \, t_{\text{spont}} \, v_0^2}{c^3 \, g(v_0) \, t_{\text{photon}}} \tag{3.40}$$

This is the threshold population inversion required for oscillation near the line $g(v_0)$. For minimum threshold inversion the largest value of $g(v)$ has been taken which corresponds, of course, to the line centre v_0.

A number of significant conclusions concerning the choice of suitable laser materials can be drawn from this important equation, but first it is necessary to express $g(v_0)$ more explicitly.

The shape of the curve $g(v_0)$ will depend on the physical processes which give rise to it. These processes can be divided into two groups, in which the line broadening mechanism is termed inhomogenous or homogenous.

Inhomogenous broadening happens because a particular frequency on the $g(v)$ curve can be associated directly with a particular atom in the laser medium. If the medium is a gas then the broadening is caused by motion of the atoms—each atom 'sees' a radiation field of different frequency depending on the velocity of its motion. For a solid state material inhomogenous broadening is caused by the close proximity of atoms whose strong electric fields cause the energy levels associated with each atom to be disturbed and thus giving a spread in linewidth.

Homogenous broadening, on the other hand, is a result of probability. For instance, pressure broadening results in a line of finite width because the presence of other atoms results in collisions which have the effect of altering the lengths of the wave trains undergoing transitions. The collisions are a random process so it is not possible to associate a given frequency with a given atom—only a probability can be specified. Thus, unlike the case of inhomogenous broadening removal of a particular atom in the system will cause a lowering in the overall height of the $g(v)$ curve rather than leave a gap at a particular frequency which is what happens with an inhomogenously broadened line. As, by definition, the area under the $g(v)$ curve is always normalized to unity the linewidth will narrow slightly. As more and more atoms are removed the linewidth will get smaller and smaller as is to be expected because the probability of one atom colliding with another becomes less.

Inhomogenous and homogenous broadening are characterized by lines, i.e. $g(v)$ curves, of different shapes. An inhomogenously broadened line has Gaussian shape and a homogenously broadened line has a Lorentzian shape. The equations of these curves are given below and their relative shapes indicated in fig. 3.2.

For a Gaussian curve :

$$g(v) = \frac{2 \, (\pi \, ln \, 2)^{\frac{1}{2}}}{\pi \, \Delta v} \exp\left[-\left(\frac{v - v_0}{\Delta v/_2}\right)^2\right] \tag{3.41}$$

41

$$\therefore \qquad g(\nu_0) = \frac{2(\pi ln2)^{\frac{1}{2}}}{\pi \Delta \nu} \qquad\qquad (3.42)$$

$$\Delta \nu = \frac{2\nu_0}{c}\sqrt{\left(\frac{2kT ln2}{m}\right)} = 7 \cdot 162 \times 10^{-7} \times \nu_0 \sqrt{\frac{T}{m}} \quad (3.43)$$

where m is the mass of the atom.

For a Lorentzian curve :

$$g(\nu) = \frac{\Delta \nu}{2\pi[(\nu - \nu_0)^2 + (\Delta \nu/2)]^2} \qquad\qquad (3.44)$$

$$\therefore \qquad g(\nu_0) = \frac{2}{\pi \Delta \nu} \qquad\qquad (3.45)$$

$$\Delta \nu = \frac{1}{\pi \tau} \qquad\qquad (3.46)$$

where τ is the time interval between each collision.

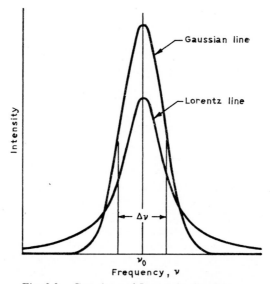

Fig. 3.2.   Gaussian and Lorentzian lineshapes.

Substituting the value of $g(\nu_0)$ into the threshold equation 3.40 for the Lorentzian line (it is not necessary to consider the Gaussian value as from equations 3.42 and 3.45 it can be seen to be a constant multiple of the Lorentzian) we have :

$$N_2 - N_1 \frac{g_2}{g_1} \geqslant \frac{4\pi^2 \nu^2 \Delta \nu}{c^3}\left(\frac{t_{spont}}{t_{photon}}\right) = \Delta N_c \qquad (3.47)$$

where ΔN_c refers to the critical or threshold population inversion.

For the smallest possible population inversion it can be seen from equation 3.47 that the following requirements for suitable laser media are :

(1) A material must be selected in which the lower level population N_1 can be kept small. A fast relaxation rate out of the level to other levels is necessary.

(2) The atomic linewidth $\Delta \nu$ should be as small as possible. Cooling of the laser material will clearly be an advantage here.

(3) The spontaneous emission time t_{spont} out of the upper level should be kept as short as possible but that out of the lower level to other levels should also be short, and not greater than t_{spont}, otherwise the lower level will become ' clogged up ' and population inversion will be impossible.

(4) t_{photon} should be as long as possible ; this means that the output mirror should be of high reflectivity and losses in medium and the cavity (due to optically inhomogenous materials, poor quality mirrors etc.) should be as small as possible.

(5) In the case of the Lorentzian line, $\Delta \nu$ is independent of frequency, as indicated in equation 3.46, but inversely dependent on t, the time between collisions. This implies that high gas pressures should be avoided otherwise $\Delta \nu$ will be excessively large. The minimum population inversion is proportional to ν^2.

In the Gaussian (inhomogenous) case, the linewidth is proportional to the frequency, cf. equation 3.43, so that on substitution for $g(\nu_0)$ from equation 3.42 in equation 3.40 it can be seen that the minimum population inversion is proportional to ν^3 which implies that the prospects for construction of X-ray or γ-ray lasers are poor. The temperature T should also be kept low to reduce $\Delta \nu$.

3.4. Minimum Pumping Power

It is instructive to get some quantitative idea of how much population inversion must be obtained for three and four-level lasers respectively. Energy levels representing these two types are shown in fig. 3.3.

(a) Four-level case.

Here $E \gg kT$ and hence levels 1, 2 and 3 start off effectively empty, hence only a few atoms need be pumped into level 2 to achieve an inversion with respect to level 1. For the threshold of oscillation N_2 must be equal to ΔN_c, the critical population inversion.

It follows therefore that :

$$P_{min} = \frac{\Delta N_c \, h\nu}{t_{spont}} \qquad (3.48)$$

43

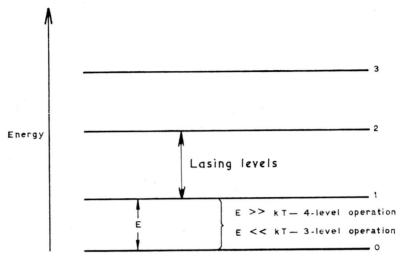

Fig. 3.3.   Comparison of three and four-level lasers.

Substituting for ΔN_c from equation 3.47 we have :

$$P_{\min} = \left[\frac{4\pi^2\nu^2\Delta\nu}{c^3} \left(\frac{t_{\text{spont}}}{t_{\text{photon}}} \right) \right] \frac{h\nu}{t_{\text{spont}}} \tag{3.49}$$

$$= \frac{4\pi^2 h\nu^3\Delta\nu}{c^3 t_{\text{photon}}}. \tag{3.50}$$

(b) Three-level case.

Referring to fig. 3.3, in this case $E \ll kT$ and the population of level 1 is about half the total number of atoms present : the other half being in level 0.   Therefore to satisfy the threshold condition it is necessary that $N_2 - N_1 = \Delta N_c$ and so the required population inversion for a three-level laser must be much greater that that for a four-level laser by a factor $\dfrac{\Delta N_c + N_1}{\Delta N_c}$ i.e. $1 + \dfrac{N_1}{\Delta N_c}$.   Normally N_1 is very much greater than ΔN_c so the factor becomes $\dfrac{N_1}{\Delta N_c}$ or $\dfrac{N}{2\Delta N_c}$ where N denotes the total number of atoms present, as N_1 is roughly half the total number of atoms present.

As a natural example of the difference between a three-level and four-level laser, consider the case of a typical solid-state laser where, for a cubic centimetre of laser material, ΔN_c is 10^{16} and $\frac{1}{2}N$ is about 10^{18}. For a four-level system the threshold condition is $N_2 = \Delta N_c$ or 0.5% of the total number of atoms.   For a three-level system, however, the

44

critical value of N_2 will be $10^{18} + 10^{16} \simeq 10^{18}$ i.e. for a four-level laser approximately 10^{16} atoms will have to be pumped into level 2 and for a three-level laser the number required will be approximately 10^{18} indicating that for a three-level laser the pumping power will have to be about a hundred times greater.

CHAPTER 4

resonators, mirrors and modes

4.1. *Resonators*

GAIN in a laser is increased by placing the active laser medium in between two mirrors which face each other. This arrangement of mirrors was familiar long before the advent of lasers and is known as a Fabry-Perot interferometer or a Fabry-Perot etalon. Consider what happens when a population inversion exists within the active medium and a spontaneous transition produces a photon which travels along the axis of the system. This photon can then interact with an atom in an excited state to give stimulated emission and a wave, which increases in amplitude, passes through the active medium and out to one of the mirrors. The amplification, or gain, of the wave is further increased by reflection backwards and forwards within the laser cavity by means of the mirrors. A laser without any mirrors can act as an amplifier ; with mirrors it becomes an oscillator. In each case a necessary condition for operation is that a population inversion occurs within the active medium. Any oscillator is essentially a device which returns some of the output from a system back into the input, such that the returned energy is in phase with the input energy. Oscillators can be mechanical, electrical or, as in the case of the laser, optical. The feedback in a laser is of course achieved by the mirrors.

In order that some energy can be obtained as an output, one of the mirrors of a laser is made slightly transmitting and is hence known as the output mirror. The optimum transmission of the output mirror depends on the gain of the laser. If the transmission is too low, very little or no output will be achieved. If the transmission is too high the number of passes up and down the cavity may be insufficient to provide enough gain to overcome the losses. As a rough guide 1 or 2% transmission may be optimum for a low gain laser line, while 10% or more may be needed if the gain is very high.

4.2. *Laser Mirrors*

The only desirable source of loss in a laser is the transmission of the output mirror. Absorption and scattering at the mirrors must be kept as low as possible and ideally, one mirror of the laser should be 100% reflecting. Conventional mirrors made of aluminium or silver are inadequate for most types of laser as their absorption is high and very high reflectivities are impossible to obtain.

Laser mirrors are almost always made by the successive deposition of high and low dielectric materials onto substrates of the highest optical quality.

The theory of multilayer mirrors can be explained as follows: suppose a ray of light of unit amplitude is propagating through a medium of refractive index n_1 when it strikes the plane surface of a material of refractive index n_2 at an angle α to the normal. Then refraction will take place at an angle β and reflection at an angle α as indicated in fig. 4.1.

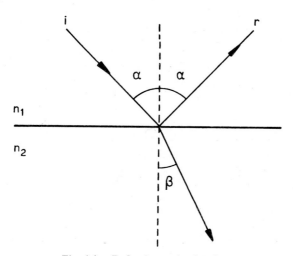

Fig. 4.1.   Reflection and refraction.

The amplitude of the reflected ray r can be shown to be of the following form[2]:

$$r = -\frac{n_2 \cos \beta - n_1 \cos \alpha}{n_2 \cos \beta + n_1 \cos \alpha}. \tag{4.1}$$

For normal incidence the intensity of the reflected light R is obtained by putting $\alpha = \beta = 0$ in equation 4.1 and squaring

$$R = \left(\frac{n_2 - n_1}{n_2 + n_1}\right)^2 \tag{4.2}$$

if $n_2 > n_1$ then r becomes negative i.e. the wave changes in phase by π.

If two parallel surfaces are considered, then if the ray reflected from the first surface is made to be in phase with the ray reflected from the second surface the two reflected rays will constructively interfere to increase the reflected intensity. Some light will be transmitted but in a multilayer structure successive surfaces can reflect more and more light back to reinforce the first reflection. By making the optical thickness of each layer a quarter of a wavelength thick very high reflectivities can be achieved. The limit to the number of layers possible is mainly due to absorption within the materials, problems of adhesion and difficulties in accurately monitoring thickness.

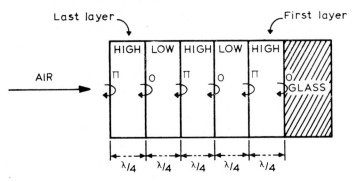

Fig. 4.2. Phase changes at the dielectric boundaries of a multilayer mirror.

Figure 4.2 is a schematic diagram of a 5-layer mirror. Some consideration will show that if the last layer to be deposited is made of high refractive index material the first reflection will suffer a phase change of π. As a result, if the next to last layer is of lower refractive index and $\lambda/4$ thick, (λ refers to the wavelength *in the material*) the first two reflected rays will be in phase.

The first layer to be deposited on the substrate is normally always of high refractive index material as it is much easier to monitor the quarter wavelength thickness by observing a minimum in intensity transmission than a maximum. As the last layer must always be of high refractive index material it follows that all such mirrors consist of an odd number of layers starting and finishing with one of high refractive index.

Silica substrates are normally used and for the highest reflectivities which exceed 99% as many as 21 to 25 layers may be deposited. The actual dielectric materials used vary according as to whether soft or hard coatings are used.

Soft coatings are made of alternate layers of zinc sulphide ZnS (refractive index = 2·2 to 2·3) and cryolite Na_3AlF_6 (refractive index = 1·38). Hard coatings may consist of layers of titanium dioxide TiO_2 (refractive index = 2·2) and silicon monoxide SiO (refractive index = 1·75 to 1·8).

The first stage in coating is to thoroughly clean the mirror substrate. This improves adhesion and reduces scattering. A procedure which works well is to gently rub the surface of the substrate with liquid detergent using a lint-free cloth. The substrate is then thoroughly rinsed in tap water, ultrasonically bathed in deionized water and analar grade acetone. Finally a 10 minute vapour bath in iso-propyl alcohol is given.

48

The materials are then placed in boats within the evaporator while the substrates are supported above the boats in a calotte. The boats are usually made of molybdenum, tungsten or tantalum. Molybdenum has the advantages of having a high melting point, is easily obtained vacuum clean and can be worked to any desired shape.

After pumping down to between $10^{-5} - 10^{-6}$ torr the boats are heated in turn to deposit the appropriate layers of material. Heating is usually indirect (i.e. by a filament around the boat) when making soft coatings as there is less tendency for violent eruptions of material in the boat. Zinc sulphide and cryolite both sublime so indirect heating provides a sufficiently high temperature. For hard coatings some materials do not sublime and must be evaporated from the melt. Hence the boats have to be heated directly in order to achieve high enough temperatures. Direct heating is achieved by passing an electric current through the boat.

As their name implies hard coatings are resistant to damage, however there are difficulties in deposition. Titanium dioxide readily dissociates to form other titanium oxides which absorb visible wavelengths. For this reason hard coatings are usually made by evaporating in an oxygen atmosphere to prevent dissociation. Further, all air must be excluded as the titanium readily reacts with nitrogen to form a nitride which is also highly absorbing. The high temperatures required for hard coatings can result in evaporation of the material forming the boat and the compounds of molybdenum and tungsten so formed are absorbing, consequently some form of external heating would be advantageous. Experiments with a carbon dioxide[20] laser show that the high temperatures produced by the beam might provide a satisfactory alternative.

Thickness monitoring is accomplished by using a white light source with colour filters and observing appropriate maxima and minima in transmission as each layer is deposited.

4.3. *Alignment of Laser Mirrors*

Laser mirrors are said to be aligned when imaginary lines drawn normal to the centre of each mirror are coincident. However, before describing a useful method for aligning a laser it should be pointed out that mirrors and, where applicable, Brewster windows (see page 56) should be kept scrupulously clean and free from all traces of dust and grease. To this end sealed cavities are desirable and attention should be paid to the environmental conditions under which the laser is used.

Laser mirrors can be cleaned by pouring a little iso-propyl alcohol onto the mirror surface and wiping the surface with a lint-free cloth. The same piece of cloth should never be used twice.

Attention should also be paid to the back face of the output mirror substrate. While obviously not as important as the front surface a dirty back surface can cause degradation of the output.

49

Accurate alignment of the mirrors is often of paramount importance if the laser is to operate efficiently or in some situations at all. This is particularly true in the case of gas lasers where the distance between the mirrors may be more than a metre. Short cavities, which are to be found in some solid state lasers and all semiconductor lasers, do not require such critical alignment especially if the gain is high. The following alignment procedure is usually adequate if the front and back surfaces of the mirrors are reasonably parallel :

(1) The power supply is first switched off to avoid the possibility of the beam entering the eye. It must be emphasized that even the lowest power laser could cause serious eye damage if the beam were focussed onto the retina. Some further discussion of this will be found in Chapter 11.

(2) An illuminated grid is placed at one end of the laser and positioned in the manner indicated in fig. 4.3.

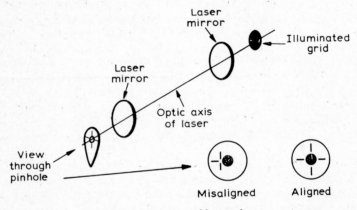

Fig. 4.3. Alignment of laser mirrors.

A suitable illuminated grid can easily be made by replacing the glass of a torch by a piece of perspex onto which a network of black lines has been drawn.

(3) A lining up card is made by drawing a black cross on a piece of white card and making a small pinhole at the centre of the cross. The operator then looks through the pinhole down the axis of the laser towards the laser mirror. If the side of the card with the cross is nearest the mirror then a reflection of the cross will be seen from the back surface of the nearest laser mirror.

(4) The operator can now see the illuminated grid and the reflection of the cross off the back surface of the mirror nearest to him.

(5) By moving the head, a position can be found where it is apparent to the operator that he is looking straight down the bore of the tube

at the grid. The mirror nearest to the operator is then adjusted until the reflection of the centre of the cross appears to coincide with the grid. When this is so the optic axis of the near mirror is pointing straight down the laser tube.

(6) The procedure is then repeated at the other end of the laser.

(7) The laser is switched on and if lasing does not occur the adjusting screws of one of the mirrors are slowly oscillated until a flash of light is seen. The mirrors are then carefully adjusted for optimum output.

The optical cavity of Javan's original laser consisted of two plane mirrors. A property of such a plane mirror configuration is that the alignment is highly critical—an accuracy of about one second of arc being necessary (the angle subtended by a golf ball at 5 miles!), consequently the output of the laser is very sensitive to vibrations, thermal changes and optical inhomogeneities within the cavity. In order to increase the alignment tolerance, one, or both, of the mirrors in all commercially available lasers is made slightly spherical. This has the effect of increasing the misalignment tolerance to only one minute of arc. Plane mirror cavities must be aligned by means of an auto-collimator as the method described above is not accurate enough. For lasers of more than several metres or so in length (e.g. carbon dioxide lasers) the use of an auto-collimator is also advisable.

It can be shown that for an optical cavity formed by two mirrors many modes of oscillation are possible. These modes can be divided into two groups : transverse modes and axial modes.

4.4. *Transverse Modes*

A beam of light propagating down the optic axis of the laser will be amplified and, provided sufficient gain is available, will emerge from the laser output mirror. Generally speaking, a photon travelling at right angles to the optic axis will not be amplified to form a laser beam as, although it may cause many stimulated emissions, there will not be a large enough number to overcome all losses, in other words the gain will be inadequate.

However in the situation where a wave is travelling only slightly off axis the wave may be able to zig-zag or ' walk ' between the mirrors a sufficient number of times to produce enough gain to overcome the losses. The result of this is to produce an output which may consist of a variety of complicated patterns of light as shown in fig. 4.4.

These patterns are the result of the laser operating in what is referred to as transverse modes. Such modes have well known equivalents in microwaves and are designated by the same notation as TEM_{pq} modes (TEM standing for transverse electromagnetic). The laser output may consist of a mixture of different transverse modes. Figure 4.4 indicates how the modes are labelled. The most commonly used mode is the TEM_{00} mode which will be discussed in detail. A laser designed to

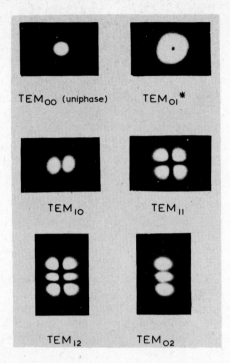

Fig. 4.4. Some low order transverse modes of a laser. The modes are generally labelled TEM_{pq} where $_p$ and $_q$ are integers referring to the number of zeros of intensity in the x and y directions respectively. The $\text{TEM}_{01}*$ mode consists of TEM_{01} modes in quadrature.

operate in the TEM_{00} mode may often be found to be oscillating in the TEM_{01}^* or 'doughnut' mode. This is often a result of a particle of dirt on the mirror surface.

The TEM_{00} is in fact the so called uniphase mode which has been referred to earlier.

It should be noted that provided the mode pattern is stable with time any transverse mode or combination of transverse modes constitute a spatially coherent output.

4.5. *The Uniphase or* TEM_{00} *Mode*

The intensity distribution across a section of the TEM_{00} mode is Gaussian and is expressed by the following equation :

$$I(x) = I_0 \exp (-2r^2/\omega^2) \tag{4.3}$$

where I_0 is the intensity at the centre (the optic axis of the laser), r is the distance from the centre and ω is the value of r for which the intensity has fallen to e^{-2} of its value at the centre. Figure 4.5 a shows the shape of the intensity distribution.

52

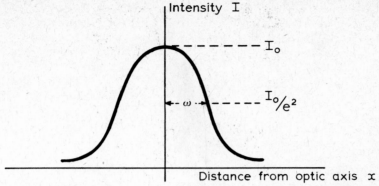

Fig. 4.5 (*a*). Gaussian intensity distribution.

Figure 4.5 *b* illustrates how the uniphase mode propagates itself through space. The two lines represent a section through the loci of the e^{-2} intensity points. It can be seen from fig. 4.5 *b* that the mode has a beam waist where a minimum value of ω, defined as ω_0, occurs. At the beam waist the wavefront is flat. For minimum diffraction loss, the wavefront curvature at a mirror must be the same as the curvature of that mirror. Thus the position of the beam waist will depend on the radii of the mirrors used. For example, if one of the laser mirrors is plane then the wavefront will be plane at that mirror. In some situations the use of plane or convex mirrors is undesirable and if a plane wavefront is required outside the cavity then passing the beam through a convex lens enables a plane wavefront to be formed at some convenient position. See fig. 4.5 *b*.

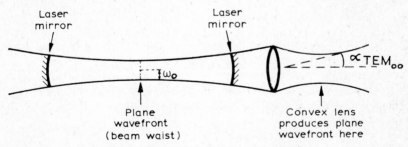

Fig. 4.5 (*b*). Propagation of TEM$_{00}$ mode.

The uniphase mode has a number of properties which often make it the most desirable mode in which to operate and consequently many lasers are designed to work exclusively in this mode.

The uniphase mode does not suffer any phase reversals across its wavefront as do higher order modes. This can be an important consideration in interferometric applications and in holography.

53

The nearest approach to a uniphase mode with a flat wavefront (i.e. at the beamwaist) is a uniformly illuminated aperture. It is interesting to compare the subsequent propagation of these intensity distributions.

In the case of the uniphase mode the semiangular divergence of the beam is given by :

$$\alpha_{TEM_{00}} = \frac{\lambda}{\pi \omega_0} \tag{4.4}$$

and the mode continues to propagate as a Gaussian distribution. Further it can be shown that $86 \cdot 5\%$ of the energy is contained within an area of radius ω.

For uniform illumination the situation is quite different. The diffraction pattern is a circular fringe system and the central bright region within the first dark ring is known as the Airy disc. This contains 84% of the energy and the first dark ring makes an angle with the optic axis given by :

$$\alpha_{uniform} = 1 \cdot 22 \frac{\lambda}{\omega_0} \tag{4.5}$$

Thus a comparison of equation (4.4) and equation (4.5) indicates that in the case of the uniphase mode more energy is contained within a smaller angle compared with uniform illumination. This is an extremely important consideration in situations where a high energy density is required by drilling, burning, welding, etc.

The following equation gives the relationship between ω, ω_0, the wavelength λ and the distance x from the beam waist :

$$\omega = \omega_0 \left(1 + \frac{x^2 \lambda^2}{\pi^2 \omega_0^4}\right) \tag{4.6}$$

Figure 4.6 shows how the TEM_{00} mode propagates by tabulating values of ω against values of x and ω_0 for a helium-neon laser operating at 6328 Å. It should be observed that the beam is virtually of constant width for some distance after which it increases its angle of divergence until the latter reaches a constant value. The small divergence of large diameter laser beams may well lead to their extensive use in space communications.

4.6. *Choice of Mirrors*

The first c.w. laser of Javan employed mirrors to actually seal off the ends of the tube containing the gas. This practice is not followed nowadays as the discharge can cause damage to the mirrors. Instead the mirrors are mounted externally and the tube is sealed by attaching two optical flats at an angle to the optical axis. The angle is that which causes minimum reflection losses for light polarized normal to the plane of the flats and is known as the Brewster angle. The flats are referred

ω_0 \ x	1 cm	10 cm	1 m	10 m	100 m	1 km	10 km	10^2 km	10^3 km	10^4 km	10^5 km	10^6 km	10^7 km	10^8 km	10^9 km	10^{10} km
0·1 mm	0·102 mm	0·225 mm	2·02 mm	2·01 cm	20·1 cm	2·01 m	20·1 m	201 m	2·01 km	20·1 km	201 km	2010 km	2·01 × 10^4 km	2·01 × 10^5 km	2·01 × 10^6 km	20·1 × 10^6 km
0·5 mm	0·5 mm	0·502 mm	0·642 mm	4·06 mm	4·03 cm	40·3 cm	4·03 m	40·3 m	403 m	4·03 km	40·3 km	403 km	4030 km	4·03 × 10^4 km	4·03 × 10^5 km	4·03 × 10^6 km
1 mm	1 mm	1 mm	1·02 mm	2·25 mm	2·02 cm	20·1 cm	2·01 m	20·1 m	201 m	2·01 km	20·1 km	201 km	2010 km	20100 km	201000 km	2·01 × 10^6 km
5 mm	5 mm	5 mm	5 mm	5·02 mm	6·42 mm	4·06 cm	40·3 cm	4·03 m	40·3 m	403 m	4·03 km	40·3 km	403 km	4030 km	40300 km	403000 km
1 cm	1 cm	1 cm	1 cm	1 cm	1·02 cm	2·25 cm	20·2 cm	2·01 m	20·1 m	201 m	2·01 km	20·1 km	201 km	2010 km	20100 km	201000 km
10 cm	10 cm	10 cm	10 cm	10 cm	10 cm	10 cm	10·2 cm	22·5 cm	2·02 m	20·1 m	201 m	2·01 km	20·1 km	201 km	2010 km	20100 km
1 m	1 m	1 m	1 m	1 m	1 m	1 m	1 m	1 m	1·02 m	2·25 m	20·2 m	201 m	2·01 km	20·1 km	201 km	2010 km
10 m	10 m	10 m	10 m	10 m	10 m	10 m	10 m	10 m	10 m	10 m	10·2 m	22·5 m	202 m	2·01 km	20·1 km	201 km
100 m	100 m	100 m	100 m	100 m	100 m	100 m	100 m	100 m	100 m	100 m	100 m	100 m	102 m	225 m	2·02 km	20·1 km

Fig. 4.6. Propagation of TEM$_{00}$ transverse mode. The beam radius, ω, at a distance x from the beam waist is given for various values of ω_0. The wavelength is 6328 Å.

to as Brewster windows. They are usually placed as shown in fig. 4.7 and not parallel, in order to avoid beam displacement.

The losses for the plane of polarization parallel to the Brewster windows are high and consequently the output of a laser using Brewster windows is always polarized normal to the plane of the windows.

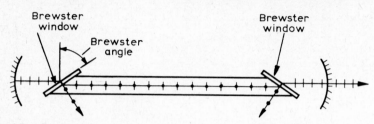

Fig. 4.7. Gas laser showing how the gas tube is terminated with Brewster windows and how the output is plane polarised.

For solid state lasers such as those using ruby or neodymium in YAG (yttrium aluminium garnet) the ends of the crystal can be polished flat and parallel and coated with materials giving the required degree of reflection. In this case no Brewster angles are necessary and the output can be unpolarized. Alternatively it may be required to use external curved mirrors and so the laser crystal has to be cut at the Brewster angle.

For reasons discussed above, the TEM_{00} mode is usually the most desirable mode in which the laser can oscillate. The achievement of oscillation in this mode and the exclusion of higher order modes is obtained by careful choice of the components of the laser.

There are three factors to be considered : cleanliness, laser tube diameter and the choice of mirrors having suitable radii of curvature.

The gain of the TEM_{00} mode is usually always greater than that of higher order modes. The output of a laser will in general consist of higher order transverse modes unless it is constrained to start operating in the uniphase mode in which case there will not be enough gain available for the higher order modes to oscillate. The uniphase mode can be made to oscillate by making the losses for the higher order modes large. This can be achieved by taking advantage of the fact that the higher modes have larger beam diameters and beam divergences and hence any stops or apertures placed within the laser cavity will increase the losses of the higher modes much more than that of the uniphase mode. It is not possible to specify precisely how large such an aperture should be as mirror quality and flatness of the Brewster windows are unknown variables which cannot be taken into account. A speck of dirt or defect on one of the mirrors, for example, could be so positioned that it was on the axis of the fundamental mode and yet it would have little effect on the TEM_{01}^{*} mode. Consequently the gain of the TEM_{00} mode would be much less than that of the TEM_{01}^{*} mode and the latter mode would oscillate.

The use of a stop within the laser cavity is not always possible particularly when the laser is designed as a sealed unit or when the mirrors are very close to the rod or tube containing the lasing medium. A better method is to use the laser tube, or rod, itself as the aperture and to choose the radii of the mirrors so that the ratio of fundamental mode diameter to the laser tube diameter can be adjusted by varying the separation of the mirrors.

This will be a convenient point to discuss the various types of resonator used. These are referred to as follows where r_1 and r_2 are the radii of curvature of the mirrors and L is the distance between the mirrors :

(a) plane parallel $r_1 = r_2 = \infty$
(b) large radius $r_1 \gg L, r_2 \gg L$
(c) confocal $r_1 + r_2 = 2L$
(d) concentric $r_1 + r_2 = L$
(e) hemispherical $r_1 = L, r_2 = \infty$

Figure 4.8 illustrates these configurations and gives an indication of how the radiation within the cavity distributes itself.

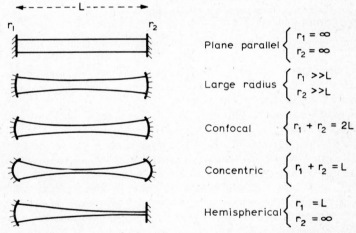

Fig. 4.8. Laser mirror configurations. The lines between the mirrors refer to the uniphase mode and are the loci of points where the intensity has fallen to i/e^2 of its value on axis. In the case of the confocal and concentric resonators these lines are drawn on the assumption that the mirrors are of equal radii.

Not any combination of mirror radii and mirror separation will sustain oscillation. These high loss and low loss combinations are indicated diagramatically in fig. 4.9.

Each type of resonator will now be considered in more detail.

(a) plane parallel.

This resonator was used in the first helium-neon laser. Fox and Li[21] have shown that the output from such a resonator is not plane but

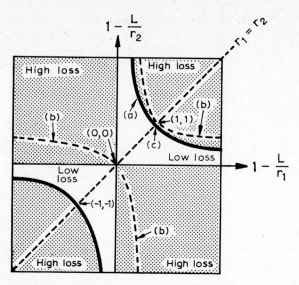

Fig. 4.9.   Loss diagram for various laser resonator configurations.   The cavity length
is denoted by L and the radii of the mirrors by r_1 and r_2.   Lines (a) and (b)
are for concentric and confocal systems respectively, (c) being the point when
two plane mirrors are used.   The shaded areas denote high loss regions.

in fact has a slight curvature due to large diffraction losses at the edges of
the mirrors.   These diffraction losses arise as a direct consequence of
the wave nature of light which causes the beam to 'spill out' over the
edges of the mirrors.   This occurs with any laser cavity irrespective of
the mirrors used.

The diffraction losses for the TEM_{00} mode can be shown to be about
half that for higher order modes whereas for a confocal resonator whose
tube diameter is roughly equal to the mode diameter the losses for the
TEM_{00} mode are $1/25$ those of the higher modes.   Thus the plane
parallel resonator is very susceptible to dirt or optical inhomogenities
such as distorted Brewster windows and can easily oscillate in higher
modes in preference to the fundamental mode.

It can be seen that if the reflectivity of the output mirror is R then on
average the beam will have to pass up and down the cavity $1/(1-R)$
times before leaving.   If the reflectivity is 99% then 100 passes will
have to be made.   Consequently unless the mirrors are very accurately
aligned the beam will 'walk off' before a sufficient number of passes
have been made.   The alignment of the mirrors of a plane parallel
resonator has to be accurate to about one second of arc and so the
resonator is not only very difficult to align but very susceptible to vibration and thermal effects which cause the output to fluctuate in intensity.
The method of alignment described earlier is not sensitive enough to
align plane mirrors.   A high quality auto-collimator is required.

Reference to fig. 4.9 shows the plane parallel configuration to lie on the boundary between a low loss and a high loss region. Any slight curvature of mirrors or Brewster windows could cause the system to enter the high loss region with consequent loss of output. Mirrors and Brewster windows have to be made flat to $\lambda/100$ in a plane parallel resonator.

The plane parallel resonator is only of historical interest as far as gas lasers are concerned. Its sole advantage lies in the fact that the mode volume is large, i.e. the oscillating mode fills up the entire cavity and thus makes maximum use of the available excited atoms within the gas tube or crystal rod. However, large outputs are not a characteristic of such systems as it can be shown that the diffraction losses are very high.

(b) large radius.

It is intuitively apparent that if the radius of curvature of the TEM$_{00}$ mode is made to coincide with that of the laser mirrors a lower loss and hence a more stable and powerful output will be obtained. The radius R of the uniphase wavefront can be shown to be given by

$$R = x\left(1 + \frac{\pi^2 \omega_0^2}{\lambda^2 x^2}\right) \tag{4.7}$$

substituting $\omega_0 = 1$ mm $\lambda = 0.5$ μm and $x = 25$ cm (the distance along the optic axis from the beam waist) gives R to be greater than 300 m. Thus virtually flat mirrors are required and would be difficult to align. In addition it is found to be extremely difficult to make mirrors having a radius of curvature more than 20 or 30 m. In practice therefore mirrors of 20 m or less radius of curvature are used. They are easier to align than plane parallel mirrors (the required accuracy of alignment is about 10 seconds) and the mode pattern they produce still fills a comparatively large volume of active medium. Lasers requiring high power outputs therefore usually use the large radius configuration.

The minimum spot size ω_0 for uniphase output obtained by using a laser having two mirrors of radii r_1 and r_2 is given by

$$\omega_0^4 = \frac{\lambda^2}{\pi^2} \frac{L(r_1 - L)(r_2 - L)(r_1 + r_2 - L)}{(r_1 + r_2 - 2L)^2} \tag{4.8}$$

the spot sizes at the two mirrors are given by

$$\omega_1^4 = \frac{\lambda^2}{\pi^2} \frac{r_1^2 L(r_2 - L)}{(r_1 - L)(r_1 + r_2 - L)} \tag{4.9}$$

and

$$\omega_2^4 = \frac{\lambda^2}{\pi^2} \frac{r_2^2 L(r_1 - L)}{(r_2 - L)(r_1 + r_2 - L)} \tag{4.10}$$

59

if the mirrors are of equal curvature

$$\omega_0^4 = \frac{\lambda^2 L (2r - L)}{4\pi^2} \tag{4.11}$$

and

$$\omega_1^4 = \omega_2^4 = \omega^4 = \frac{\lambda^2 r^2}{\pi^2} \frac{L}{(2r - L)} \tag{4.12}$$

in the case of the large radius mirror resonator $r \gg L$ and so

$$\omega^4 = \frac{\lambda^2}{\pi^2} \frac{rL}{2} \tag{4.13}$$

and so any change in L has only a very small effect on ω. For example if L is 40 cm and r is 10 m and $\lambda = 4880$ Å then ω is approximately 0·469 mm. If L is reduced to 50 cm then ω is only increased to 0·496 mm. It can be seen therefore that the large radius resonator is not very useful where fundamental mode selection is made by increasing the beam diameter.

(c) confocal.

If the radii of the two mirrors are reduced still further until their foci coincide, or, in the case of mirrors having equal radii, their radii are both equal to the mirror separation, then the resonator is said to be confocal[22,23]. As would be expected the mode volume is reduced. For a confocal resonator giving a uniphase output and having mirrors of equal radii

$$\omega^2 = \frac{r\lambda}{\pi} \tag{4.14}$$

and

$$\omega_0^2 = \frac{r\lambda}{2\pi} \tag{4.15}$$

$\therefore \omega/\omega_0 = \sqrt{2} \approx 1\cdot4$ which is still adequately large.

For some types of lasers it is desirable to keep the laser tube as small as possible and so far for a fixed mirror separation the resonator system having the smallest mode volume should be chosen. By differentiating equation 4.12 with respect to L it will be seen that the confocal system satisfies this requirement.

As with large radii mirrors the mode diameter is a slowly varying function of mirror separation. Reference to fig. 4.9 shows that when the confocal system uses mirrors of equal radii then the system is on a boundary between a low loss and high loss region and is therefore susceptible to poor quality mirrors or Brewster windows. If the confocal system is set up it may be found that the output is unstable owing to operation in the high loss region. Therefore it is essential that some adjustment of mirror separation should be provided in order that operation in the stable low loss region is ensured.

The confocal resonator is by far the easiest to align, an accuracy of only $1\frac{1}{2}$ minutes of arc being required. It also has maximum discrimination against higher order modes : the gain for the TEM_{00} mode is at least 25 times greater than that for higher order modes.

(d) concentric.

Concentric systems are formed by reducing the radii of the mirrors still further until their sum equals the mirror separation. If the mirrors are of equal radius then the mirror separation is equal to twice the radius. This system is referred to as spherical.

In general, concentric systems are as difficult to align as a plane mirror system, have a large mode diameter and lie on the boundary between high and low loss regions. For these reasons concentric resonators are seldom used for gas lasers although, as is explained later, they do have advantages in solid state systems.

(e) hemispherical.

The hemispherical resonator is made up of a spherical mirror and a plane mirror placed at its centre of curvature. This results in a mode having a large diameter at the spherical mirror and a spot, limited in its smallness by diffraction, at the plane mirror.

By putting $r_2 = \infty$ into equations 4.9 and 4.10 the following expressions for the uniphase beam radii are obtained

$$\omega_1^4 = \frac{\lambda^2}{\pi^2} \frac{r_1^2 L}{r_1 - L} \qquad (4.16)$$

at the spherical mirror.

$$\omega_0^4 = \omega_2^4 = \frac{\lambda^2}{\pi^2} L(r_1 - L) \qquad (4.17)$$

at the plane mirror.

The difference in the spot sizes at the mirrors is evident from these equations, clearly r_1 can never be quite equal to L otherwise the beam becomes infinite in diameter at the curved mirror and of zero diameter at the plane mirror. Figure 4.9 also shows that if $L > r_1$ the system is working under high loss conditions. Thus in practice the mirror separation is made slightly less than the radius of the curved mirror. As an example suppose $\lambda = 4880$ Å and $L = 50$ cm and we require ω_1 to be 1 mm then $r_1 - L$ is about 3 mm.

From equation 4.16 it can be seen that a small change in mirror separation will have a considerable effect on the value of ω_1. The hemispherical system is therefore to be recommended where elimination of higher order modes by mode diameter variation is required.

As is apparent from fig. 4.8 the fundamental transverse mode is almost cone shaped so that only about 1/3 of the volume between the mirrors is utilized, and hence is not as efficient in utilizing the active material of the laser as a large radius mirror resonator. This results in a lower power output, perhaps about half that obtainable from the large radius mirror resonator.

The alignment sensitivity of the hemispherical resonator is about equal to that of the confocal system, i.e. it is relatively easy to align. This is obvious because misaligning the plane mirror will only result in a smaller area of the curved mirror being used. Care must be taken to ensure that the axis of the curved mirror is pointing down the centre of the laser tube. This is achieved by replacing the plane mirror by a curved one so that the system becomes confocal. After alignment the curved mirror is removed and a plane mirror substituted so that the mirror separation is less than the mirror radius. The plane mirror is aligned and then moved away from the curved mirror ; the resultant increase in beam diameter at the curved mirror gradually eliminates the higher order transverse modes until only the fundamental transverse mode oscillates.

Having described the various types of resonator some explanation of the types usually employed in gas and solid state lasers is now possible.

(i) gas lasers.

Where the highest possible output power is required a large radius mirror resonator is used. This makes maximum use of the active laser medium consistent with easy alignment and freedom from intensity fluctuations. Uniphase operation is ensured by careful choice of mirror separation and tube diameter, by avoiding dirt through appropriate sealing, and by using mirrors and Brewster windows of the highest optical quality.

Where power considerations are not so important and uniphase operation under adverse environmental conditions is essential, the hemispherical system is frequently used.

When high gain is required and the type of laser is such that this is obtained by using the smallest possible tube diameter then the confocal system is preferable.

(ii) solid state lasers.

Here a distinction must be drawn between high power pulsed operation and low power c.w. operation. In the former case the power densities involved are considerable and can easily damage optical coatings. Thus the first consideration in cavity design for high power pulsed lasers is the avoidance of damage from small spot sizes at optical components.

Solid state lasers are generally much shorter in length than gas lasers and therefore tend to support the oscillation of higher order transverse modes. In ruby lasers optical inhomogeneities due to crystal defects and temperature variations make the uniphase mode difficult to obtain and can even result in the output being spatially incoherent.

Where maximum power output is needed, either c.w. or pulsed, and where higher order modes are not undesirable, a plane mirror resonator

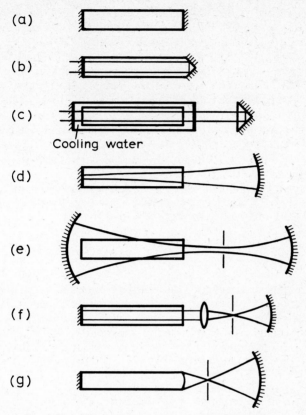

(a)

(b)

(c)

Cooling water

(d)

(e)

(f)

(g)

Fig. 4.10. Some solid state laser cavity configurations.

is used. Difficulties in maintaining alignment are overcome by polishing the ends of the rod flat and parallel and coating them with silver as in fig. 4.10 *a*.

Alternatively the end of the rod may be prism shaped as in fig. 4.10 *b* so as to effectively double the length of the rod.

It may be necessary to place additional optical components such as Q-switches inside the laser cavity. In this situation a mirror external to the laser rod must be employed. In one commercial system the total internal prism is a separate component and two sapphire windows enclose the laser rod and act as the ends of a water jacket, see fig. 4.10 *c*. The cooling water reduces the possibility of damaging the rod.

In order to ensure that c.w. solid state lasers operate in the TEM_{00} mode a good arrangement is a near hemispherical system with one end of the rod flat and silvered and the other end coated to minimize reflection owing to the high refractive index of the material (fig. 4.10 *d*). The spherical mirror is pulled away from the rod until the losses for all the higher order modes are so great that only a uniphase output is obtained. The relative insensitivity of the hemispherical resonator to misalignment is an advantage in that mechanical vibration introduced as a result of passing cooling water round the laser rod has very little effect on the output. The hemispherical resonator is not suitable for high power pulsed solid state lasers as the small spot size can cause the mirror on the end of the rod and even the rod itself to be damaged.

The concentric system shown in fig. 4.10 e is often used. An aperture is placed as indicated at the beam waist and is most effective at removing higher order transverse modes. These stops are easily burnt and consequently are often made of diamond. However they can easily be replaced.

The arrangement shown in fig. 4.10 f is an improvement as maximum use is made of the active laser material while fig. 4.10 g shows how the lens can be eliminated by shaping the end of the rod.

4.7. *Axial Modes*

In the previous two sections we have been primarily concerned with transverse modes which manifest themselves as distributions of intensity in a transverse direction with respect to the axis of the laser. The two mirrors of the laser form a resonator and the waves within the resonator form standing wave patterns with the mirrors as nodes in an exactly analogous way to a sound wave within an organ pipe or a microwave inside a cavity. The stationary waves satisfy the relation

$$n\lambda/2 = L \tag{4.18}$$

or
$$\nu = \frac{nc}{2L} \tag{4.19}$$

where n is an integer, c is the velocity of light in the laser medium, λ is the wavelength and L is the length of the cavity, i.e. the distance between mirrors. The latter are assumed to be plane. For spherical mirrors the relation is not quite correct.

As an example of how many half wavelengths that can be fitted inside a laser cavity suppose that $L = 50$ cm and $\lambda = 0.5$ μm, then $n = 2$ million. As n has such a large value at optical wavelengths it is clear that a large number of different values of n are possible with only a very small change in wavelength. These differing values of n define modes of oscillation in the cavity which are called axial modes, each axial mode being labelled as TEM_{pqn}. For the n-1 *th* axial mode we have

$$(n\text{-}1)\frac{\lambda}{2} = L \tag{4.20}$$

By differentiating equation 4.19 with respect to n it can be shown that the difference in frequency between the two adjacent axial modes, $\Delta\nu$, is given by

$$\Delta v = \frac{c}{2L} \qquad (4.21)$$

for example :
if $L = 50$ cm then $\Delta v = 300$ MHz
if $L = 10$ cm then $\Delta v = 1500$ MHz

As equation 4.21 is independent of n, the frequency separation of adjacent axial modes must be the same irrespective of their actual frequencies. Thus the modes of oscillation of a laser cavity will consist of a very large number of frequencies, each given by equation 4.19 for different values of n and separated from each other by a frequency difference given by equation 4.21. Figure 4.11 indicates the situation.

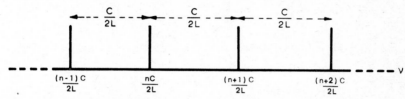

Fig. 4.11. Axial modes in a plane mirror cavity of length L.

In some applications of lasers such as interferometry and holography the axial mode content of the output can be of great importance. The gain curve of the laser output, which is the reciprocal of the laser material's absorption curve, is usually much wider than the axial mode separation and so at even moderate levels of gain more than one axial mode can oscillate. The effect of more than one oscillating axial mode is to greatly increase the linewidth of the output and so reduce coherence length. The decrease in coherence length in changing from single axial mode to multimode operation is much greater than that arising from, for instance, an increase from two to three or seven to ten axial modes. This is because the width of an individual axial mode is extremely small as it is inversely proportional to the Q of the cavity which in turn can be extremely large.

In the case of a laser oscillating in a single axial mode the coherence length at any instant in time will be very large but the linewidth can still appear considerable over a longer time period. This happens because unless the cavity length is strictly controlled the absolute wave length of oscillation will wander, thus effectively producing a large linewidth. In general single axial mode output lasers must be stabilized so as to define the wavelength as accurately as possible.

Figure 4.12 shows the gain curve which for a gas laser is the Doppler curve together with the axial modes. It will be explained later why the laser output in one axial mode is so much narrower than the natural linewidth and the cavity (Fabry Perot) linewidth. Two positions of the

65

E

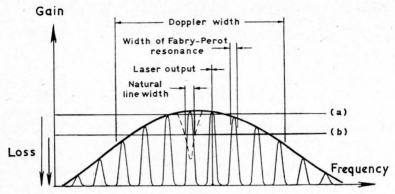

Fig. 4.12.   Axial modes and linewidths.

loss line are shown.   At level (a) only two axial modes oscillate but on increasing the gain so the loss line drops to level (b) six axial modes oscillate.

Equation 4.18 is generally only true if the laser has plane mirrors and is operating in the uniphase TEM_{00} mode.

For the more general case where the laser cavity consists of spherical mirrors and higher order transverse modes can operate, the resonance condition for a TEM_{pqn} mode is given by[23] :

$$\left[n+\frac{1}{\pi}(1+p+q)\cos^{-1}\left(1-\frac{L}{r}\right)\right]\frac{\lambda}{2} = L \qquad (4.22)$$

where r is the mirror radius (assuming both laser mirrors have the same radius) and L is the cavity length.

For the uniphase mode $p = q = 0$ and for plane mirrors $r = \infty$ and so equation 4.22 reduces to equation 4.18.

From equation 4.22 it is apparent that each transverse mode has a number of axial modes associated with it of differing frequencies and separated by $c/2L$ irrespective of the value of p and q.

An interesting situation arises if

$$\cos^{-1}\left(1-\frac{L}{r}\right) = \frac{\pi}{2} \qquad (4.23)$$

in which case

$$\left[n+\frac{1+p+q}{2}\right]\frac{\lambda}{2} = L \qquad (4.24)$$

Equation 4.23 implies that $r = L$, i.e. a confocal resonator.   It can be seen from equation 4.24 that if $p + q$ is odd, the frequency of the transverse mode will fall exactly on the frequency of another uniphase mode of different value of n.   On the other hand, if $p + q$ is even, the

66

higher order mode will fall half way between two uniphase modes. Consequently if a confocal resonator is operating in the uniphase mode, the axial modes will be separated in frequency by $c/2L$ and if in higher order modes, by $c/4L$.

It has been explained that the coherence length of the laser depends on the number of oscillating axial modes in the output. In order to describe this dependence it is useful to consider the laser output being used in a Michelson interferometer. The visibility V of the fringes obtained is defined by

$$V = \frac{I_{max} - I_{min}}{I_{max} + I_{min}} \qquad (4.25)$$

where I_{max} is the intensity at the centre of a bright fringe and I_{min} is the intensity at the centre of a dark fringe. It can be shown that for N axial modes of equal intensity the dependence of visibility on path difference in the arms of the interferometer is as shown in fig. 4.13.

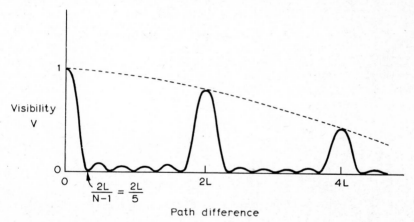

Fig. 4.13. Visibility curve resulting from laser oscillating in N-6 axial modes of equal intensity and separation and of finite linewidth.

The gradual overall decrease in visibility indicated by the broken line arises from the finite width of each axial mode of which more will be said later. It will be noted that regions of good visibility appear every even multiple of the length of the laser tube and that if N modes oscillate the coherence length $\mathscr{L}$ is given by

$$\mathscr{L} = 2\frac{2L}{N-1} \qquad (4.26)$$

Thus the smaller we can make N the longer will be the coherence length. If only one axial mode oscillates the increase in coherence length is enormous and in practice is determined by the Q of the cavity and ultimately by noise. It should be emphasized that the curve shown in fig. 4.13 represents that applicable to an ideal situation. In practice the axial modes vary in amplitude and their individual amplitudes vary in time. In addition modes which oscillate can cause the gain for a neighbouring mode to be insufficient and so entirely prevent its oscillation.

An important factor affecting the number of oscillating axial modes is the width of the gain curve of the laser material. The wider the linewidth the more axial modes will be supported. Some approximate values of these linewidths are given in fig. 4.14.

Ruby	300,000 MHz
Nd-glass	300 Å
Nd-CaWO$_4$ Nd-YAG	15 Å
Semiconductor	cooled, 50 MHz. room temperature, 1–20 Å according to current.
He-Ne	1700 MHz
He–Cd	3000 MHz
Ar	3500 MHz
Kr	3500 MHz
CO$_2$	50 MHz
liquid	usually 50–100 Å, with grating mirror 1 Å, with grating mirror and etalon 10^{-2}Å.

Fig. 4.14. Approximate gain curve widths for various types of laser.

One way of obtaining single axial mode operation is to reduce the cavity length so that the mode separation is greater than the linewidth. For a helium-neon laser if L is made less than 10 cm only one axial mode will oscillate. In the case of the ruby laser a cavity length of a few centimetres will ensure single axial mode oscillation.

However a better method is to use a coupled cavity technique. This is achieved by either adding a third mirror outside the laser cavity[24] or inserting an optical flat into the cavity[25,26]. Fig. 4.15 a shows the gain curve of the laser and the axial modes of the laser resonator.

In each case we are effectively superimposing another set of axial modes indicated in fig. 4.15 b on those due to the laser cavity. If the second cavity is made sufficiently short then the possible modes will be separated by a greater amount than the linewidth of the laser medium and only one axial mode will oscillate as in the case in fig. 4.15 c.

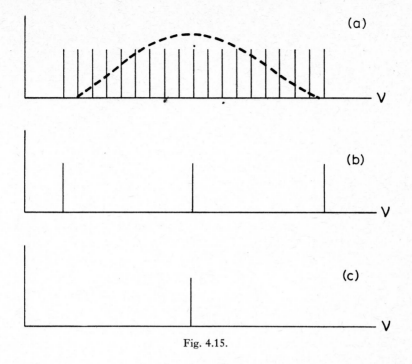

Fig. 4.15.

The output from a single axial mode laser will normally fluctuate in intensity because of variations in the cavity length which cause the mode to drift along the gain curve. This is undesirable in most applications and a method for overcoming this is described in Chapter 5.

It has been explained that the laser cavity is essentially the well known Fabry-Perot resonator. Why then is the linewidth of a laser working in a single axial mode intrinsically so much smaller than that of the output of a Fabry-Perot used as a filter? The answer to this question can be obtained by considering the quality factor or Q factor of the resonator. The Q factor can be expressed in two ways :

$$Q = \frac{\nu}{\Delta \nu} \tag{4.27}$$

$$Q = 2\pi\nu\beta/\alpha \tag{4.28}$$

where β = energy stored in the resonator, where α = energy lost per second from the resonator, where ν is the frequency of oscillation and where $\Delta\nu$ is the linewidth of the output.

An electrical oscillator will have a typical Q value of 100. Suppose we consider a Fabry-Perot resonator consisting of two mirrors L cm apart with one mirror 100% reflecting and the other having a reflectivity of R. Then the principle source of energy loss is on reflection. Absorption and diffraction losses can be ignored in comparison. On average, a photon will make $1/1-R$ passes before escaping, i.e. it travels a distance $2L/1-R$ at the speed of light c so the time it spends in the cavity is given by $2L/c(1-R)$; this results in a rate of loss of energy α given by

$$\alpha = \frac{h\nu c(1-R)}{2L} \tag{4.29}$$

This is the loss of energy from one photon. The energy stored within the resonator by one photon is given by

$$\beta = h\nu \tag{4.30}$$

and thus substituting equations 4.29 and 4.30 into equation 4.28 we obtain the Q of the cavity

$$Q = 2\pi\nu \frac{h\nu}{h\nu c(1-R)} 2L \tag{4.31}$$

$$= \frac{4\pi L}{\lambda(1-R)} \tag{4.32}$$

If $L = 1$ m $\lambda = 0.6328$ μm and $R = 0.95$ then Q is about 4×10^8 which by substituting into equation 4.27 gives $\Delta\nu = 1$ MHz. Thus the linewidth is much narrower than the axial mode separation and the Q very much larger than an electrical resonator. However this analysis applies to a *passive* Fabry-Perot resonator. In the case of the laser the material within the resonator actually supplies energy to the oscillating modes so that in theory the denominator of equation 4.28 can be zero and an infinitely narrow linewidth obtained. In practice loss of energy always occurs as a result of spontaneous emission contributing to the output. Even so, a linewidth of 1 Hz has been obtained from a helium–neon laser.

In this chapter we consider how some properties of the laser output can be varied, including increasing the power output by Q-switching, modulation, deflection, frequency doubling, and stabilization of the cavity to control the axial mode content of the output.

As so many methods of controlling laser beams are based on the electro-optic effect, it will be useful to describe some of the more important electro-optic effects and electro-optic materials often found in laser applications.

5.1. *The Electro-Optic Effect*

The basic characteristic of all electro and magneto-optic effects is an induced change in the refractive index of a material—usually a solid crystalline substance, but sometimes a liquid—with an applied electric or magnetic field. In many cases the change in refractive index is dependent on the polarization of the incident light—such materials are then said to be birefringent under the action of the field. Birefringence is also referred to as double refraction.

5.1.1. *The Kerr Effect*

When a material is under the action of an electric field an optic axis may be induced parallel to the direction of the field. This effect is known as the Kerr effect and occurs in all of the 32 types of crystal classes.

An optic axis may be defined as that direction in a crystal in which, when light is propagated in the same direction, the refractive index is independent of the direction of polarization of the light.

If a plane polarized light wave is incident normally on the surface of a crystal possessing an optic axis and where the optic axis is not normal to the face of the crystal, birefringence occurs. The lines drawn across the face of the crystal in fig. 5.1 indicate the direction of the optic axis. For simplicity the optic axis is considered to be parallel to the face upon which the light is normally incident. The plane polarized light on entering the crystal can be resolved into two components. One component is parallel to the optic axis and is termed the ordinary component or ordinary ray, the other is perpendicular to the optic axis and is termed the extraordinary component or extraordinary ray. The velocities of propagation of the two components within the crystal are different. If, as is the case with calcite, the velocity of the extraordinary

71

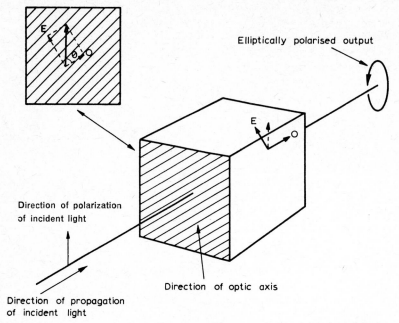

Fig. 5.1. Production of elliptically polarized light from plane polarised light by a uniaxial crystal.

ray is greater than the velocity of the ordinary ray the crystal is said to be negative. Crystals such as quartz where the velocity of the extraordinary ray is less than the velocity of the ordinary ray are said to be positive. On emerging from the crystal the ordinary and extraordinary components are, in general, out of phase and hence the output light is elliptically polarized. Three special cases are of interest : if the ordinary and extraordinary rays differ in phase by 0, 2π, 4π, etc. the light is plane polarized in the same direction as that incident on the crystal. If, however, the phase difference is π, 3π, 5π, etc. then a little thought will show that the emergent light is plane polarized at an angle 2θ to the original direction. Finally suppose $\theta = 45°$, then the amplitudes of the ordinary and extraordinary rays will be the same and if the phase difference introduced by the passage through the crystal is some multiple of $\pi/2$ the emergent light will be circularly polarized.

Some crystals are naturally birefringent, i.e. they have a naturally occurring optic axis. In fact most naturally birefringent crystals have two optic axes and are therefore called biaxial.

In the Kerr effect the applied electric field produces an optic axis, or an additional optic axis in the case of crystals which are already uniaxial or biaxial. The amount of phase retardation between the ordinary and extraordinary components is, among other things, proportional to the square of the applied voltage. Consequently the Kerr effect is also known as the quadratic electro-optic effect.

Figure 5.2 shows a voltage V being applied to electrodes of separation d fixed to the faces of a crystal. The length of crystal through which the light propagates is l. The phase retardation produced is given by

$$\Delta\phi = \frac{2\pi j l V^2 \lambda}{d^2} \tag{5.1}$$

where j is the Kerr coefficient and λ is the wavelength of light used. V is in esu volts (1 esu volt = 300 volts).

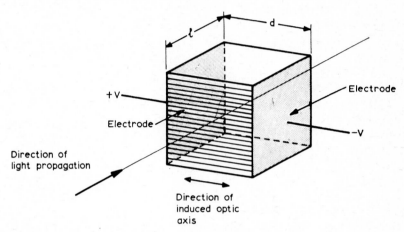

Fig. 5.2. The Kerr effect.

The Kerr constant for most crystals is very small. However for nitrobenzene the coefficient has a value of 220 which is sufficiently high to be utilized for electro-optic devices. Nitrobenzene must be extremely pure to maintain a high Kerr coefficient. As it is also toxic and unstable, most electro-optic devices used in conjunction with lasers make use of the Pockels effect which occurs in sufficient magnitude in more satisfactory materials.

5.1.2. The Pockels Effect

The Pockels effect is similar to the Kerr effect. The induced phase shift depends directly on the voltage hence the effect is often referred to as the linear electro-optic effect.

The Pockels effect can be used in two modes—as indicated in fig. 5.3. When the electrode configuration is such that the applied field is normal to the direction of propagation of the incident light, i.e. the same as in the Kerr effect, the transverse Pockels effect is said to occur, (see fig. 5.3a). When the applied field and the propagation direction are parallel, as in fig. 5.3*b* the longitudinal Pockels effect is said to take place.

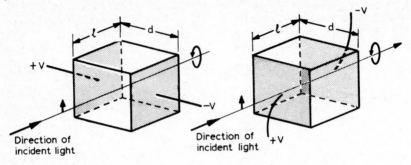

(a) Transverse Pockels effect   (b) Longitudinal Pockels effect

Fig. 5.3. (*a*) Transverse and (*b*) longitudinal Pockels effect.

Unlike the Kerr effect only non-centrosymmetric crystals are capable of exhibiting the Pockels effect—this restricts the number of crystal classes available to 21.

Using the notation of fig. 5.3 where l is the length of the crystal and d the width, the expressions for the induced phase shift are as follows: for the transverse effect:

$$\Delta\phi = \frac{\pi l n_0^3 r V}{\lambda d} \text{ radians} \tag{5.2}$$

and for the longitudinal effect where $d \equiv l$

$$\Delta\phi = \frac{2\pi n_0^3 r V}{\lambda} \text{ radians} \tag{5.3}$$

n_0 is the refractive index of the ordinary ray, r is the electro-optic constant, V is the voltage and λ is the wavelength of the light.

It can be seen from equation 5.3 that in the case of the longitudinal effect the phase shift is independent of the crystal length. This is advantageous because, in the transverse effect, inhomogenities in the crystal can result in an emergent wavefront which varies considerably in phase owing to non-uniform retardation. Consequently most Pockels effect devices use the longitudinal effect to avoid this.

An important criterion for an electro-optic material is the half-wave voltage which, as its name implies, is the voltage necessary to produce a retardation of half a wavelength between the ordinary and extra-ordinary rays. Figure 5.4 tabulates the half-wave voltages for a number of materials. In addition the electro-optic constants are also listed together with some remarks on the materials, several of which are still undergoing development. The relative magnitudes of dispersion and birefringence are of crucial importance in frequency doubling. This will be explained later in this chapter.

Material	Chemical Formula	Electro-optic const ($\mu m/V$)	Half-wave Voltage	Remarks
ammonium dihydrogen phosphate (ADP)	$NH_4 H_2 PO_4$	$24 \cdot 5\ 10^{-6}$	10600	All available as large crystals of very high quality. Hygroscopic. Use restricted to radiation of wavelength less than $1 \cdot 7 \mu m$. Transparent in the ultraviolet.
potassium dihydrogen phosphate (KDP)	$KH_2 PO_4$	$10 \cdot 5\ 10^{-6}$	8000	
potassium dideuterium phosphate (KDP)	$KD_2 PO_4$	$26 \cdot 4\ 10^{-6}$	3000	
quartz	SiO_2	$0 \cdot 2\ 10^{-6}$	30000	Can be used in the far infrared.
lithium tantalate	$LiTaO_3$	$21 \cdot 7\ 10^{-6}$	2500	Negligible birefringence. Suffers from optical damage, which can be reduced by annealing at 600°C.
lithium niobate	$LiNbO_3$	$18 \cdot 0\ 10^{-6}$	2900	Large birefringence but suffers from optical damage above 160°C. Can be used out to 4 μm.
barium strontium niobate	$Sr_{0 \cdot 75} Ba_{0 \cdot 25} Nb_2 O_6$	1380	50	Negligible birefringence. Does not suffer from optical damage. Very difficult to make.
barium sodium niobate	$Ba_2 Na Nb_5 O_{15}$		1500	Large birefringence. Does not suffer from optical damage. Very difficult to make.
lithium potassium niobate	$K_6 Li_4 NbO_3$	130	930	
strontium potassium niobate	$KSr_2 Nb_5 O_{15}$		400	
hexamine	$(CH_2)_6 N_4$	$4 \cdot 18\ 10^{-6}$	14900	
potassium tantalum niobate (KTN)	$KTa_{0 \cdot 65} Nb_{0 \cdot 35} O_3$		300	Mixed crystal. Difficult to grow well.
zinc telluride	ZnTe		2700	Large transmission range 0·4–20 μm. Hygroscopic. All cubic crystals and so none have natural birefringence.
cuprous chloride	CuCl	$6 \cdot 1\ 10^{-6}$	7200	
zinc sulphide	ZnS	$2\ 10^{-6}$	10400	
zinc selenide	ZnSe	$1 \cdot 6\ 10^{-6}$	7100	
barium titanate	$BaTiO_3$	10^{-4}	1760	

Fig. 5.4. Some electro-optic materials.

5.1.3. *The Faraday Effect*

When a magnetic field is applied to many solids, gases and liquids, the plane of polarization of an incident beam of light is rotated. If the angle of rotation is denoted by θ and the magnetic field in the direction of the light path by H, then

$$\theta = VlH \qquad (5.4)$$

where V is a constant called Verdets constant and l is the path length within the material. A material often employed in Faraday effect devices is dense flint glass which has a Verdet constant of about 25 radians cm^{-1} gauss^{-1}.

For applications involving fast switching the Faraday effect is not suitable as the time taken to switch on the magnetic field and for it to build up to its required strength is unduly long.

Sometimes it is necessary to ensure that the output from a laser is not reflected back into the laser cavity which can cause an undesirable modulation of the output. In this, and in any situation where an optical isolator is required, the Faraday effect can be employed by making use of the property that the direction of polarization of light reflected back is rotated in the same direction as that of light travelling in the forward direction. If, at the input end, the light is vertically polarized and the magnetic field so adjusted to rotate the plane of polarization through 45°, then a polarizer at the output end orientated at 45° to the vertical will pass the light. However, any light subsequently reflected back through the system will have its direction of polarization rotated a further 45° and will thus be at 90° to the orientation of the input polarizer. The light will therefore be prevented from passing further back.

Figure 5.5. shows how such an optical isolator works.

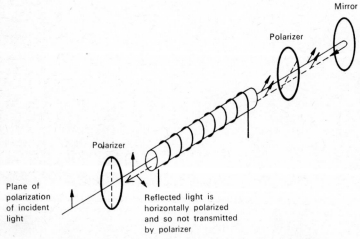

Fig. 5.5. The optical isolator based on the Faraday effect.

76

5.2. Q-switching

The power output of many lasers can be greatly increased by a process known as Q-switching. For lasers which are normally pulsed, a single high power pulse of shorter duration can be produced while continuously operating lasers can be made to give a train of pulses. For many applications such as drilling, welding, high speed photography and optical radar, high power outputs, or short pulse duration times, are needed. The distinction between power and energy must be appreciated. Q-switched lasers always give *lower* energy outputs compared with when they are not Q-switched because of absorption and other sources of loss in the Q-switch itself. Nevertheless the power output is very much higher because the pulse duration is shorter. Equation 5.5 gives the elementary relation between power output, energy in the pulse and pulse duration.

$$\text{power output (watts)} = \frac{\text{pulse energy (joules)}}{\text{pulse duration (seconds)}} \qquad (5.5)$$

In the case of the ruby laser for example the pulse energy is typically 10 joules delivered in one millisecond. Q-switching reduces the pulse duration to 10 nanoseconds and so increases the power output from one kilowatt to one gigawatt (10^9 watts).

Q-switching, as its name implies, involves changing the Q of the laser cavity so that feedback by the mirrors is suppressed and so depletion of the population of the upper laser energy level is not permitted until its population has reached a high value[27]. The laser is pumped with the resonator kept at a very low Q value ; the Q of the cavity is then made to suddenly increase allowing gain from stimulated emission to take place, with the result that the laser energy escapes in a very short, highly intense pulse. For obvious reasons, this process is also known as Q-spoiling or giant-pulse operation.

Q-switching is carried out by placing a closed shutter, which may take various forms, in the cavity and so effectively eliminating the resonator from the laser medium. After the laser is pumped the shutter is very rapidly opened so restoring the cavity to the system.

There are two important requirements to be met if Q-switching is to be possible. These are :

(a) the rate of pumping must be faster than the spontaneous decay rate from the upper energy level. If the pumping is too slow, the upper level will empty faster than it can be filled and sufficient population inversion will not be achieved.

(b) the Q-switch must switch rapidly in comparison with the build up in stimulated emission, otherwise the latter will be a gradual process and a longer pulse time than necessary will be obtained so reducing the power. In practice a Q-switch time of at least 10 ns is desirable.

There are many ways of attaining Q-switched operation. The following methods work well and are to be found in common use.

5.2.1. *Rotating mirror method*

If one of the mirrors forming the laser cavity is rapidly rotated in the manner shown in fig. 5.6 then only for the instant when the mirror is correctly aligned with respect to the optic axis will a high Q exist. Synchronization of the pumping flash with rotation of the mirror is clearly necessary. This type of switch is compact, reliable, cheap and simple and has the advantage that it can be used in conjunction with a laser operating on any wavelength even in the infrared. It does, however, tend to be slow in switching as 60,000 r.p.m. is about the highest rotation rate that can be achieved using air bearings and driving the mirror or prism by means of an air turbine.

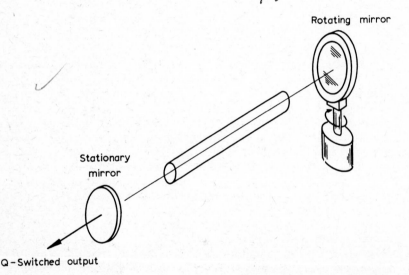

Fig. 5.6.   Rotating mirror Q-switch.

Figure 5.7 is a photograph of a rotating mirror Q-switch used on a carbon dioxide laser and driven by an electric motor at up to 40,000 r.p.m.

The fact that one of the mirrors forming a laser cavity can be rotated slightly without the laser output ceasing implies that, in the rotating mirror Q-switch, there will be some finite time during which the gain exceeds the loss. This time may be sufficiently long so that, in each rotation, a Q-switch output is obtained which actually consists of a train of pulses, in the case of the ruby laser, each one lasting 20–50 ns. Depending on the degree of pumping, there may be as many as a dozen pulses in each train. The power in each pulse is reduced as the total

Fig. 5.7. Rotating Mirror Q-switch.
(S.E.R.L. photograph).

available energy must be shared amongst others in the train. Single
pulse operation can be achieved by incorporating a component in the
cavity known as a Lummer-Gehrke plate in which case the device is
known ·as a Daly-Sims[28] Q-switch. This arrangement is shown in
fig. 5.8.

79

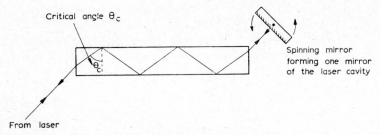

Fig. 5.8.   The Daly-Sims Q-switch.

The sides of the plate are optically flat and parallel.   If the angle θ is made equal to the critical angle then the entire system is more sensitive to slight deviations of the rotating mirror away from the normal to the axial ray shown.   This sensitivity can be increased by making the plate long and so increasing the number of total internal reflections.   In this way the switching time can be decreased and single phase operation obtained.

5.2.2. *The Electro-Optic Q-switch*

A very fast Q-switch can be made using the Kerr[29] or Pockels effect described at the beginning of this chapter.

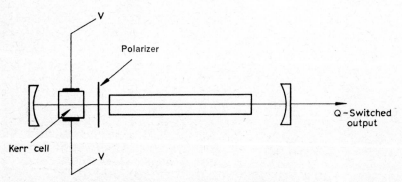

Fig. 5.9.   Electro-optic Q-switch using the Kerr effect.

If the output from the laser is not naturally polarized a polarizer is placed in the cavity together with an electro-optic crystal (see fig. 5.9) maintained at such a voltage that the plane polarized light incident on the crystal is converted into circularly polarized light by transmission through the crystal.   The laser mirror reflects this circularly polarized light and in so doing reverses the direction of polarization.   Hence on re-emerging from the electro-optic crystal the light is again plane polarized but at 90° to its original direction.   It is therefore not transmitted by the polarizer.   When the voltage across the plates is reduced to

80

zero, usually done by shorting them out, the electro-optic crystal has no effect other than a small source of loss within the cavity.

The change in voltage which must obviously be synchronized with the pumping can be accomplished in less than 10 ns and so very effective Q-switching takes place.

Voltages required for Kerr effect operation are usually higher than for the Pockels effect. In addition the undesirable properties of nitrobenzene and carbon disulphide, which are among the few materials having a sufficiently high Kerr constant, have led to commercial electro-optic Q-switches being based on the Pockels effect in ADP (ammonium dihydrogen phosphate) or KDP (potassium dihydrogen phosphate). Unlike nitrobenzene these materials are solids but they need to be hermetically sealed or placed in a suitable liquid as they are hygroscopic.

The electrodes on the faces of the crystal in longitudinal effect Pockels cell Q-switches operating in the longitudinal mode are usually in the form of thick metal rings. Uniform metallic layers would suffer damage due to the high power densities. The high electrical resistance of such thin films also prevents fast switching.

Fig. 5.10. Neodymium-YAG laser showing the Pockels cell Q-switch and polariser on the left of the cavity and a lithium niobate frequency doubler on the right. (Courtesy Laser Associates).

F

Figure 5.10 shows a neodymium laser. Next to the left hand mirror is a Pockels cell and polarizer forming the Q-switch. A lithium niobate frequency doubler is also included next to the right hand mirror.

5.2.3. *Photochemical* Q-*switching*

A photochemical substance, such as a saturable dye[30,31], forms a very convenient Q-switch ; it has the great advantage that pumping synchronization is not required. A thin film of dye is placed in the cavity and when the intensity builds up to a sufficiently high level the dye is. rapidly bleached, increasing the Q of the cavity, and a giant pulse is emitted. A popular dye is a 1–2 mg l^{-1} concentration of cryptocyanine solution which is contained in a small cell as in fig. 5.11 and placed in the cavity at the Brewster angle. Uranium doped glass is also used. Both of these materials can be used over and over again without replacement as the bleaching action is only temporary.

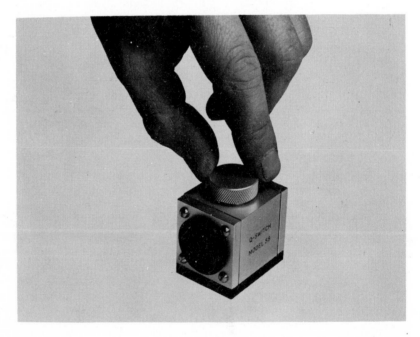

Fig. 5.11. A photochemical Q-switch cell containing a saturable dye.
(Courtesy Laser Associates).

5.2.4. *Exploding Film* Q-*switches*

Figure 5.12 shows a simple method of Q-switching which like the saturable dye Q-switch requires no pump synchronization. A thin film of absorbing material is supported on a glass plate and placed between two lenses also in the cavity so that the laser beam comes to a focus at the absorbing film. When the population inversion is sufficiently high the power density at the focus is high enough to explode the material and cause Q-switching to take place[32,33]. Aluminium is a suitable material for the absorbing layer.

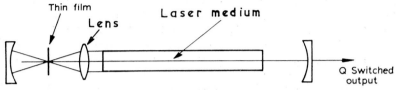

Fig. 5.12. Exploding film Q-switch.

With exploding film Q-switches the output is single-shot and a fresh piece of film must be suitably positioned before the next Q-switched pulse can occur.

5.3. *Modulation of the Laser Output*
Amplitude modulation of the laser output is important in communication applications and considerable efforts have been made to find methods of obtaining high frequency modulation. The following methods have all proved successful although the electro-optic modulator appears to be the only type which will enable the wide bandwidth capabilities of optical communications systems to be realized.

5.3.1. *Internal Modulation*
The easiest and most obvious way to modulate the output of a laser is to modulate the pump power.

For helium-neon lasers[34] the highest modulation frequencies are limited to about 100 kHz on account of the rise time of the oscillator and the finite time required to obtain an excited neon atom by resonant energy transfer from an excited helium atom. The relation between pump power and output is very non-linear, therefore for linear modulation the percentage modulation of the output cannot be very high.

Semiconductor diode lasers on the other hand can be modulated at frequencies up to 10^{10} Hz by modulating the excited current.

5.3.2. *Mechanical Modulators*
External modulation can be achieved by feeding the output of a laser into a Michelson or Twyman-Green interferometer. If the path length in one of the arms is varied the output will be modulated accordingly. Path length variation is obtained by attaching one of the mirrors to a piezo-electric crystal[35] which has the property of changing its size under the action of an applied electric field. Thus modulation of the voltage

across the crystal produces a corresponding modulation in the intensity of the output beam.

Modulation frequencies of up to 5 MHz with a high percentage of modulation are possible.

5.3.3. *Acoustic Modulation*

Under the action of a compressional acoustic wave some materials suffer a periodic change in refractive index. When the acoustic wavelength is of the order of the wavelength of light the material behaves as a diffraction grating so that an incident laser beam is split into a number of orders[36]. In liquids this is known as the Debye-Sears[37] effect and in solids as the Raman-Nath[38,39] effect.

There are basically two ways in which these effects can be exploited for modulation.

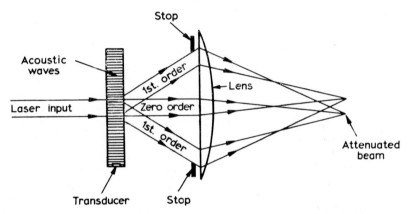

Fig. 5.13. Acoustic modulation by attenuation of first order diffracted beams.

If the laser beam is incident parallel to the plane of the acoustic fringes, plus one and minus one order diffraction beams can be obtained, as indicated in fig. 5.13, together with a directly transmitted zero order beam. By appropriate placing of an aperture these first order beams can be truncated by equal amounts. The second lens recombines the attenuated first order beams with the unaffected zero order beam to produce a plane wavefront of different intensity to that emerging from the acoustic grating. By varying the acoustic frequency, the angle of diffraction can be changed and hence the degree of attenuation of the first order beams which in turn modulates the output.

An alternative approach[41] is to allow the laser beam to fall on the acoustic grating at an angle as shown in fig. 5.14. When the angle θ is related to the acoustic (λ_a) and optical (λ_0) wavelengths by the following equation

$$\lambda_0 = 2\lambda_a \sin \theta \qquad (5.6)$$

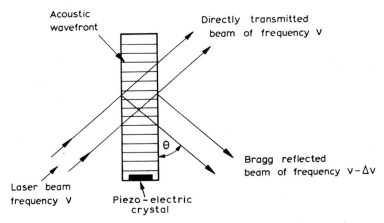

Fig. 5.14. Acoustic modulation using reflection at the Bragg angle.

Bragg reflection is said to occur in which case the fringe planes act as partially reflecting mirrors and consequently a large percentage of the incident light is reflected. As the acoustic grating is effectively moving perpendicular to the fringe direction the reflected wave is reduced in frequency by the frequency of the acoustic wave by virtue of the Doppler effect. The reflected and transmitted waves are then recombined to produce a carrier and a single side band. This single sideband operation is often preferred as it is more economical in bandwidth requirements.

Water and fused silica respectively are often used in liquid and solid modulators while a piezo-electric transducer is used to generate the acoustic wave.

5.3.4 *Absorption Modulators*

There are a number of ways in which solid state absorption modulators can work. The absorption edge modulator[40] will be described here.

The transmissivity of many solid materials varies with the wavelength of the incident light. Fairly sharp changes in transmissivity with wavelength occur which are known as absorption edges. The position of these absorption edges with respect to wavelength are also found to vary with the magnitude of an applied voltage. Consequently if a laser beam whose wavelength corresponds to an absorption edge is directed at the material the transmission can be altered by altering the voltage across the material as indicated in fig. 5.15, i.e. the transmitted light can be modulated.

Unfortunately this method has serious disadvantages in that the light induces photoconductivity which in conjunction with the large fields required results in high power dissipation and consequent cooling problems within the device.

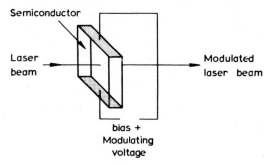

Fig. 5.15. Absorption modulation of a laser.

5.3.5. *Electro-optic Modulation*

Apart from pumping modulation, all the modulation methods described so far involve external modulation of the laser output. Electro-optic methods can be used either internally or externally and, by virtue of the rapid way in which electro-optic devices respond to variations in applied voltage, they probably offer the most promising methods of modulation.

Internal modulation[42] using the electro-optic effect can be accomplished by means of an arrangement depicted in fig. 5.16.

A longitudinally excited Pockels cell is placed within the laser cavity and, as explained earlier, serves to rotate the plane of polarization of reflected light on application of a voltage. Thus the light passing through the Pockels cell from left to right and undergoing reflection at the laser mirror is further rotated in polarization on re-passing through the cell. If the plane of polarization has been rotated through 90° a suitably placed polarizing device acts to reflect this light out of the cavity at an angle. In a gas laser the polarizing device can be the Brewster window itself.

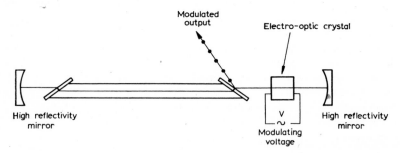

Fig. 5.16. Electro-optic internal modulation of the output of a gas laser.

With this system the percentage modulation and efficiency can be very high and linear, broad band modulation can also be achieved.

Lasers can also be electro-optically modulated externally by a longitudinally excited Pockels[43] cell, this arrangement probably being the most popular method of modulation. Figure 5.17 shows the mode of operation. If the laser does not produce a polarized beam, two crossed polarizers are placed outside the laser cavity with the Pockels cell in between. The Pockels cell is so oriented that its two optic axes are at 45° to the polarization of the laser beam entering the cell. Therefore the light leaving the cell can be considered as four components, one pair polarized

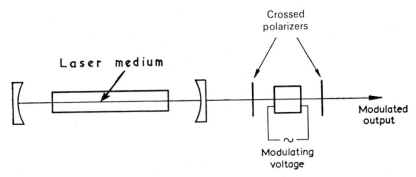

Fig. 5.17. Arrangement of components for external electro-optic modulation of a laser.

at right angles to the plane of polarization of the second polarizer and the other pair polarized in the same direction. The former pair will be blocked and the latter pair will be transmitted. Interference will then take place between the transmitted pair and the resultant intensity will depend on the relative phase difference introduced between them on their passage through the cell. If they are π, 3π, 5π, etc. out of phase destructive interference will occur and the light intensity will be zero. On the other hand if the phase difference is 0, 2π, 4π, etc. maximum intensity will result.

The intensity of the output can be varied by changing the voltage on the cell. Generally the phase difference introduced is $2n\pi + \alpha$ where n is an integer and so the intensity transmitted I can be shown to be related to the incident intensity as follows:

$$I = I_0 \sin^2 (\alpha/2) \qquad (5.7)$$

A plot of I/I_0 against α is shown in fig. 5.18 (a) and it can be seen that a linear region of operation occurs near $\alpha = \pi/2$. As linear operation is obviously desirable the voltage on the cell can be biased to ensure that a phase difference of $\pi/2$ holds when the amplitude of the modulating voltage is zero. See fig. 5.18 (b).

An alternative method of optical biasing is to insert a quarter wave plate between the first polarization and the cell. Suitably orientated this produces a phase change of $\pi/2$ before the light enters the cell.

87

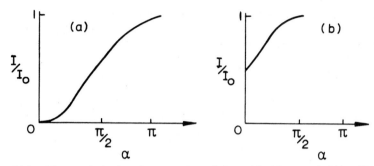

Fig. 5.18.   Characteristic of an electro-optic modulator (a) without optical bias (b) with $\lambda/4$ optical bias.

The ratio of maximum to minimum transmission of the modulator is known as the extinction ratio and there are many reasons why high values of this ratio are not achieved.   If the beam diverges the natural birefringence of the crystal will result in phase differences being produced. Similarly any lack of flatness or parallelness and any inhomogeneities within the crystal or in the applied electric field can also reduce the extinction ratio.   In practice an extinction ratio of 30:1 is just acceptable.

Most modulators use ADP or KDP as the crystalline material.   These are easily obtained as large crystals of good optical quality.   Cubic crystals such as zinc telluride and zinc selenide have no natural birefringence and so are preferable when convergent light must be used. Other electro-optic materials such as lithium niobate, lithium tantalate and barium titanate have much higher electro-optic coefficients but are not always easily obtainable as large pure crystals and all are very temperature dependent so stabilization must be provided.

Unlike Pockels cell Q-switches, the electrodes are not ring shaped as power densities are usually not high enough to cause problems.   The electrodes are often either metallic or metallic oxides evaporated onto the crystal faces with optical windows laminated onto the faces to produce a sealed unit.   Tin (II) oxide (SnO), cadmium oxide (CdO) and indium oxide (InO) are often used for electrodes.   The lower the resistance of the electrode the higher the frequency response.   Low resistance electrodes are obtained by depositing thick layers.   However these reduce the light transmission so a suitable balance between requirements must be maintained.

The upper frequency limit of electro-optic modulators is about 100 KHz.

5.4. Deflection of the Laser Output

There are several applications of lasers in which the beam from a laser must be deflected.   The deflection can either be a regular scanning or deflection from one position to another.   Mechanical methods using rotating mirrors can be used but are slow and the large masses, which are often involved, restrict their use to scanning. The most promising beam deflection systems use electro-optic or acoustic devices and in each case depend on refractive index changes to alter the direction of propagation of an incident light beam[44].

5.4.1. *Electro-optic Beam Deflection*

Digital beam deflection in which the beam is directed to one of a large number of discrete points can be achieved by using a combination of electro-optic and birefringent crystals[45]. Figure 5.19 indicates the basic principles of the process.

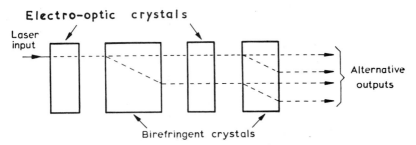

Fig. 5.19.   A two-stage electro-optic beam deflection system.

Figure 5.19 shows a two stage deflector in which the laser beam can be deflected to one of four colinear positions. The first component can be a Pockels cell which can be switched off or to a predetermined voltage and which rotates the plane of the light emerging from the cell through $\pi/2$ with respect to its incident polarization. The second crystal can be of calcite which is a birefringent material and which has its optic axis so directed that the beam is deflected to one of two positions depending on the direction of polarization of the incident beam. This combination of electro-optic cell and birefringent crystal constitute the basic unit of the system. Many of these units are placed in tandem until the required number of final beam positions are obtained. Using n such units 2^n positions can be addressed.

The most important features of such a system are the resolution, the total number of positions and the switching time.

The resolution or smallness of the deflected spot is greatly enhanced by using convergent light which comes to a focus on the final address plane. However there is a limit on this size because large angles of convergence resulting in very small diffraction spot sizes cannot be tolerated because of the large amount of unwanted background light produced. Further, this background light is also the limiting factor in the maximum number of digital positions which can be addressed. Deflectors capable of addressing the array of 256×256 positions have been built using 16 stages.

Switching times are governed by the same factors which control those in modulators except that thick electrodes of low resistance are even less tolerable as the combined effect over many stages can act to greatly reduce the output intensity. For this reason deflection rates of 100 kHz form an upper limit at the present time.

Beam scanning and analogue deflection systems are, in principle, simpler as deviation on passage through an electro-optic prism can be used to scan the beam or deflect it to any required position. However unlike digital deflectors the crystals must be very large and of uniform quality and these requirements have severely limited development so far.

5.4.2. *Acoustic Beam Deflectors*

In section 5.3.3 it was explained how an acoustic wave in a material can act as a diffraction grating. The diffracted light in general consists of a directly transmitted, or zero order beam and two first order beams symmetrically positioned on each side of the zero order beam. Now the sine of the angle which either first order beam makes with the zero order beam is inversely proportional to the grating spacing. Thus by varying the frequency of the acoustic wave the angle may be altered and the beam can address a number of positions[46,47,48]. The range over which the beam can be deflected is higher than for electro-optical deflectors, the limiting factor being the range of acoustic frequencies available. High scanning rates, in excess of 10 MHz, can be achieved.

5.5. *Frequency Doubling*

The very high power densities made available by lasers have enabled several phenomena to be relatively easily observed which previously could be regarded as theoretical curiosities. These are the so-called non-linear effects. Among the most interesting of these is frequency doubling in which radiation of frequency ν, on propagating through some crystalline materials, emerges as radiation consisting of a mixture of two frequencies, the original frequency ν and a new frequency 2ν. The double frequency component has, of course, a wavelength which is half that of the incident radiation.

The explanation of such a non-linear effect lies in the way in which a beam of light propagates through a solid material. The latter consists of atoms whose nucleus and associated electrons form electronic dipoles. Electromagnetic radiation in the form of a light beam interacts with these dipoles and causes them to oscillate which, by the classical laws of physics, results in the dipoles themselves acting as sources of electromagnetic radiation. When the amplitude of vibration of the dipoles is small the radiation they emit by virtue of their oscillation is the same as the incident radiation. However, as the intensity of the incident radiation increases so does the amplitude of vibration of the dipole. For small amplitudes there is a linear relation between these parameters but for larger amplitudes non-linearities become apparent. This results in the frequency of oscillation of the dipole having harmonics, the second and strongest harmonic being at twice the frequency of the incident radiation. For this reason frequency doubling is also referred to as second harmonic generation.

If the strength of the dipoles (actually the polarization or dipole moment per unit volume) is defined by P, and an applied electric field by E, then it may be shown that

$$P = \chi_1 E + \chi_2 E^2 + \chi_3 E^3 + \ldots \text{ etc.} \tag{5.8}$$

where χ is called the polarizability.

It follows therefore that if E takes the form of an oscillating field $E = E_0 \sin(\omega t)$, such as that produced by an electromagnetic wave, then substitution into Equation 5.8 gives

$$P = \chi_1 E_0 \sin(\omega t) + \tfrac{1}{2}\chi_2 E_0^2(1 - \cos 2\omega t) + \ldots \text{ etc.} \tag{5.9}$$

Equation 5.9 contains a term in 2ω and which corresponds to an electromagnetic wave having double the frequency of the incident wave.

The term $\chi_2 E^2$ in Equation 5.8 will not approach the term $\chi_1 E$ unless E has a value of at least 10^5 V cm^{-1} which corresponds, at optical wavelengths, to a power density of about 10^7 cm^{-2}. As the electric field produced by sunlight is approximately 10 V cm^{-1} it is not surprising that the observation of non-linear effects had to await the advent of the laser.

Not all solids exhibit frequency doubling. If the material has a centre of symmetry then an applied electric field must produce polarizations of the same magnitude but of opposite sign according as to whether the applied field is positive or negative. Consequently for materials which have a centre of symmetry such as glasses and liquids the coefficients χ_2, χ_4, ... etc. must all be zero. As χ_2 is zero, no frequency doubling can be observed.

Frequency doubling was first observed by Franken[49] and his co-workers in 1961. They focussed the 6943 Å output from a pulsed ruby laser onto a quartz crystal and obtained second harmonic generation at 3472 Å.

Early experiments in frequency doubling gave very low conversion efficiencies of about 1%. This was found to be due to the fact that dispersion within the crystal causes the frequency doubled light to travel at a different velocity to the light whose frequency is not doubled. As the latter is generating the former throughout its passage through the crystal destructive interference occurs and the frequency doubled light undergoes periodic fluctuations in intensity through the crystal. If, however, the speed of propagation of the two beams can be made the same this effect would not occur[50,51], and a much more powerful frequency doubled wave is obtained. This equalization of speeds is known as phase

91

matching and can be achieved using birefringent crystals providing the dispersion is less than the birefringence. ADP and KDP mentioned earlier fall into this category and are commonly used for second harmonic generation in commercial laser systems where efficiencies of 20–30% are available[52]. These efficiencies are high enough to make focussing of the laser beam unnecessary.

Several new materials promise even higher conversion efficiencies. Among these is lithium niobate which gives a high conversion efficiency but whose refractive index is very dependent on laser power. This effect is known as optical damage and, in the case of lithium niobate, is found not to occur above 160°C. Consequently lithium niobate frequency doublers have to be situated in an oven which must be at an accurately controlled temperature for phase matching. While ADP and KDP have a very much higher optical damage threshold, their conversion frequency is lower. A new material, barium sodium niobate, has an even higher conversion efficiency than lithium niobate and does not appear to suffer from optical damage. However it is very difficult to grow crystals of sufficient quality and is therefore very expensive. Figure 5.20 is a photograph of a barium sodium niobate crystal converting the 1.06 μm output from a neodymium laser to 0·53 μm green light. The infrared is not normally visible to the eye so a phosphor screen was used when taking the photograph. By placing the crystal inside the laser cavity, 50% conversion efficiency has been obtained[53].

5.6. *Frequency Stabilization of the Laser Output*

The axial mode content of a laser output can be controlled by adjustment of cavity length or by inserting an etalon into the cavity. In each case single axial mode output can be obtained which results in the linewidth of the output dramatically decreasing and thus increasing the coherence length.

For some applications in measurement and communications, single axial mode output, while necessary, is not sufficient. There are basically two reasons for this, both arising from variations in the length of the cavity due to vibration, temperature changes and other external factors. First the varying cavity length causes the position of the axial mode to move and so the frequency of the output varies correspondingly. Secondly this variation in frequency takes place under the gain curve (the Doppler curve in the case of gas lasers) and so the output will also vary in intensity. If the frequency separation of the axial modes is sufficiently large then even under high gain conditions the output could fall to zero.

These undesirable effects are removed by a method which is commercially available on some helium neon lasers[54].

The gain curve of a gas laser is inhomogenously broadened. This results in an effect known as hole-burning in which the gain curve dips where·an axial mode is oscillating[55]. At this frequency, the population

Fig. 5.20. Frequency doubling of the 1·06 μm output of a neodymium laser by a barium strontium niobate crystal to produce green light at 0·53 μm wavelength. The infra-red beam was made visible by a phospher coated screen.
(Photograph : Copyright 1968, Bell Telephone Laboratories, Inc., Murray Hill, New Jersey U.S.A. Reprinted by kind permission of the Editor, Bell Laboratories Record).

inversion, and hence the gain, is reduced by virtue of the laser output. If a single axial mode is not oscillating at the centre of the gain curve it is observed that a dip in the gain curve occurs at a symmetrical position with respect to the peak of the gain curve. This happens because the particular resonance concerned is actually a standing wave within the laser cavity made up of two travelling waves moving in opposite directions. Thus, reducing the number of molecules able to contribute to a population inversion having velocity $+v$ also reduces the number having velocity $-v$ and so two symmetrical holes are burned. In the case where the axial mode is at the centre of the gain curve one hole only is burned. These two situations are shown in fig. 5.21.

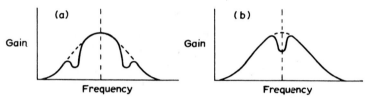

Fig. 5.21.   Hole burning (a) with axial mode not centred on the peak of the gain curve (b) with axial mode centred on the peak of the gain curve.

The dip in the gain curve when one axial mode is oscillating at the centre of the gain curve can be utilized to stabilize the frequency of the output. One mirror of the laser is mounted on a piezo-electric crystal and an alternating voltage and a d.c. bias applied to the crystal. The result is that the output of the laser will increase slightly irrespective of the direction of motion if the mode is centred on the dip. If it is not the output will increase or decrease according to the direction of motion. By using a phase sensitive detector and a suitable servo control, the output can be stabilized to better than one part in 10^7 over long periods of time.

5.7. Cavity Dumping

While Q-switching by means of a rotating mirror can produce a controlled train of output pulses up to a frequency of about 60 kHz, higher frequencies are increasingly difficult to obtain because of the mechanical problems associated with the high-speed rotation of large masses.

Trains of pulses at frequencies up to several megahertz can however be obtained from lasers capable of continuous operation by a method known as cavity dumping. In this process both mirrors of the laser are effectively made highly reflecting so that initially no radiation is emitted from the laser and so the energy within the cavity increases until the rate of the inevitable small losses balances the input power, i.e. when equilibrium is reached. At this point an optical switch is operated which couples out all the radiation previously stored within the cavity. This output takes place rapidly; in fact in the time taken for a photon to make a round trip within the cavity, i.e. $2L/c$. The optical switch is reset and the radiation within the cavity is allowed to build up again and the process is repeated.

This technique has been used successfully with both gas and solid state lasers. Neodymium–YAG lasers have had their peak powers increased by two orders of magnitude at pulse rates of several megahertz using this technique.

5.8. *Mode Locking*

Even higher rates of modulation can be achieved by placing an optical switch adjacent to one of the laser mirrors within the cavity. There will, of course, under normal lasing conditions, be a constant flow of photons out of the cavity constituting a continuous output of laser light. If the switch is now modulated at very high speed there will be imposed on the flux a periodic variation in photon density. The output will then consist of pulses of radiation which can be as short as a picosecond (10^{-12} s). The laser is then said to be mode locked. It is anticipated that such short pulses will be of great value in laser communications and in laser fusion technology.

The achievement of mode locking was the culmination of previous efforts at mode stabilization. In an unstabilized laser many axial and possibly also many transverse modes will be oscillating simultaneously and independently. The first step therefore is always to eliminate all higher order TEM modes so that only the TEM_{00} mode oscillates. The first achievement of mode locking was in 1964 at the Bell Telephone Laboratories by Hargrove and since developed by many other workers[360,361,362,363]. A modulator such as an acoustic modulator is placed within the laser cavity and driven at a frequency $c/2L$ where c is the velocity of light and L is the cavity length. This expression it will be noticed is also exactly equal to the frequency difference between adjacent axial modes. The cavity will therefore be optically homogenous twice per second. Thus light incident on the modulator at a time of zero loss will be incident on the modulator yet again (after passing up and down and undergoing one reflection) at the next occasion of zero loss. Energy which would normally be used up for modes oscillating other than that in synchronism with the zero loss condition will therefore become available for the latter mode. The oscillating axial modes are thus constrained in time, i.e. phase, to have a definite fixed phase relationship to each other. It can be shown by simple mathematical analysis that a consequence of this fixed phase relationship is that the output of the mode locked laser has the following characteristics:

    (i) The pulse repetition frequency is $c/2L$, i.e. numerically equal to the frequency separation of adjacent axial modes for a laser cavity of length L.

(ii) For N locked oscillating modes of equal amplitude the width of the pulses (in time) is $1/\Delta\nu_{osc}$, where $\Delta\nu_{osc}$ is the oscillation linewidth of the laser. This suggests that YAG and dye lasers which have large values of $\Delta\nu_{osc}$ will be most effective in producing the shortest mode locked pulses.

(iii) The peak pulse power is N times the average laser power without mode locking.

Figure 5.22 indicates some of these characteristics. Finally it should be noted that mode locking can be done by phase modulation as well as amplitude modulation described above, the properties of the output are the same.

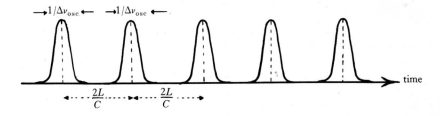

Fig. 5.22.   Mode locked output.

CHAPTER 6

solid state lasers

6.1. *The Ruby Laser*

THE first laser to be constructed used ruby as the active medium[14]. A ruby laser, when it is Q-switched, still provides a most powerful and useful source of light. The main components of the ruby laser are shown in fig. 1.11. The ruby crystal is usually 1 mm to 2 cm in diameter and between 5 cm and 20 cm long. Ruby consists of aluminium oxide Al_2O_3 to which has been added a small proportion (about 0·035%) of chromium which gives the crystal its characteristic pink colour. Sometimes higher concentrations of chromium are used in which case the crystals are red. The chromium, in the form of Cr^{3+} ions, provides the actual energy levels between which the stimulated emission, essential to laser action, takes place. Figure 6.1 shows the energy level system for Cr^{3+} in the host Al_2O_3.

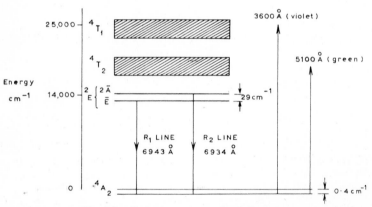

Fig. 6.1. Ruby laser energy level diagram.

The energy scale is given in cm^{-1} where $8066 \ cm^{-1} = 1 \ eV$. An explanation of the way in which the energy levels are labelled should be obtained from a textbook on spectroscopy[56], however an understanding of the nomenclature is not necessary as far as this book is concerned.

A percentage by weight of 0·035 means that there are about 10^{19} chromium ions in each cubic centimetre of crystal. A xenon flash-tube similar to a photographic electronic flash-gun is used to excite or pump the ruby ; the shape and position of the tube is most important if maximum efficiency is to be obtained. This follows because the basic three-level character of the ruby laser implies that at least half the ions in the $^4A_2$ ground level must be pumped up to the ^2E levels before laser action is possible. The flash-tube can take the form of a helical tube as in fig. 1.11, wound round the ruby rod with a coaxial cylindrical reflector. A better method is to arrange a straight flash-tube next to the ruby at the centre of the reflecting cylinder or to use an elliptical reflector with the ruby rod and flash-tube at the foci as shown in fig. 6.2[57]. Both of these arrangements are also easier to cool.

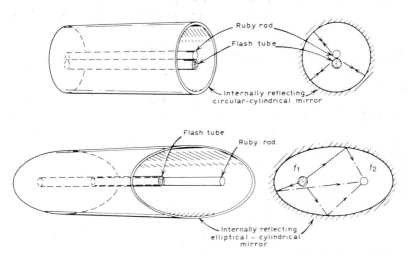

Fig. 6.2. Ruby laser flash lamp reflectors. (After Harris, *Wireless World* 1963).

The absorption spectra for ruby peaks in the violet and green regions of the spectrum[58] corresponding to the broad $^4T_1$ and $^4T_2$ levels shown in the energy level diagram for Cr^{3+} in fig. 6.1. These broad bands, each 1000 Å wide, help to make the pumping by the 'white' light from the flash-tube more efficient in comparison, for example, with pumping into a single line. The input energy to the flash-tube is provided by discharging a capacitor of 50 to 1000 μF through a tube filled with the gas xenon to a pressure of 150 torr to which has previously been applied a constant voltage of about 1 to 2 kV. The discharging capacitor triggers the breakdown of the gas, producing a flash of pumping light of typically 500 to 1000 joules in energy and lasting for about 1 ms. The laser output has a power of about a kilowatt lasting for approximately a millisecond, consequently the output energy is of the order of a joule and so the efficiency is very much less than 0·1% since most of the energy supplied by the capacitor is dissipated as heat. ′

After pumping to the absorption bands the ions drop rapidly down in less than 10^{-7} s to the very closely spaced 2E levels which are only

29 cm^{-1} (870 GHz) apart. The transition to the ground state from the E level is called the R_1 transition and corresponds to a wavelength of 6943 Å. The transition from the 2Ā level to the ground state is called the R_2 transition and the photon produced has a wavelength of 6929 Å. Under normal conditions the threshold for the R_1 line is less than that for the R_2 line. In addition the spontaneous lifetime[59] in the E_1 level is greater than the thermal relaxation time for energy transfer between the 2Ā and E levels. As a result the R_1 line oscillates in preference to the R_2 line.

The ground state actually consists of two lines of only 0·4 cm^{-1} (12 GHz) separation. At room temperatures the thermal energy of the ions results in the output being 200–300 GHz wide so the separation is not apparent. On cooling the ruby to 77°K, however, the linewidth decreases to 3 GHz and two lines are resolved. As the wavelength is also strongly temperature dependent, the R_1 output changes to 6934 Å with an overall linewidth of 15 GHz.

The ruby laser starts to lase approximately 0·5 ms after the start of the pumping flash and so lasts about half a millisecond (unless, of course, the laser is Q-switched). Once started, stimulated emission rapidly depopulates the upper lasing levels—much faster than the pumping rate can supply atoms and so the laser process has to pause and 'wait' until the population inversion is again achieved. The effect of this is to give an output which consists of a large number of spikes, each spike lasting about a microsecond as shown in fig. 6.3.

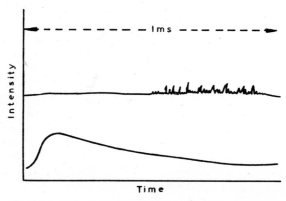

Fig. 6.3. Ruby laser and flashlamp outputs versus time.

As a consequence of the large amount of heat dissipated by the flashlamp, the laser becomes hot with the result that a limit must be set on the pulse repetition rate if overheating and damage are to be avoided. Only one pulse every few minutes is feasible if the laser is air cooled but by circulating water around the laser an increase in repetition rate to several pulses a minute is possible.

Typical commercial ruby lasers produce energy pulses from 1–100 J in about 1 ms. Continuous operation is possible but is never normally used. By Q-switching the pulse time can be reduced to 10 ns with powers of tens of gigawatts.

As has been mentioned, the linewidth at room temperatures can be as much as 300 GHz, so for a 10 cm cavity length having an axial mode separation of 3 GHz as many as 100 axial modes can be present in the output. The coherence length available will be very small, approximately one millimetre. However coherence lengths of several metres are obtainable using an etalon in the cavity[60].

As far as spatial coherence is concerned, the inevitable inhomogeneties in the ruby cause lasing to occur in individual independent filaments with little spatial coherence. Special cavities incorporating small apertures as described in Chapter 4 must be used if uniphase outputs are needed[61].

6.2. *The Neodymium Laser*

A more recently developed solid state laser than the ruby laser is the neodymium laser which is used extensively for many purposes inside and outside the laboratory. Its superiority stems from the fact that it is of the 4-level type with its lower lasing level some 2000 cm^{-1} above the ground level, i.e. it is effectively empty at room temperatures. Pumping, threshold and efficiency are therefore greatly dependent on the optical properties of the host material and while many have been investigated two have emerged as the best: glass[63,64] and YAG[65] (yttrium aluminium garnate). Calcium tungstate[66] ($CaWO_4$) has also been used with some success. The impurity ion Nd^{3+} has energy levels which are split by the presence of the host field and lasing can occur between several levels notably between those giving radiation of wavelength $0.914\,\mu m$, $1.06\,\mu m$, $1.317\,\mu m$, $1.336\,\mu m$ and $1.35\,\mu m$. However the output almost always chosen is that at $1.06\,\mu m$ with YAG as probably the most popular host material because of its high resistance to optical damage and mechanical strength. YAG also has sufficient thermal conductivity to enable high repetition rates to be used without heat dissipation becoming a problem.

The overall efficiency achievable by a Nd–YAG laser is 1–2% and such lasers have been operated continuously at 1 kW. Commercially available outputs of 100–200 W exist but for routine laboratory use and for most applications outputs of 1–20 W are more common with linewidths of 15 Å (500 GHz). The spatial coherence of such lasers is comparable with that of the ruby laser.

Nd–YAG lasers can be Q-switched, cavity dumped and mode locked as described in chapter 5. As a rule of thumb the mean pulsed power output is approximately equal to the continuous output, e.g. a 20 W c.w. laser will give by Q-switching a pulsed

99

output of 20 kW peak at a repetition rate of 10 kHz assuming each Q-switched pulse lasts 100 ns.

However, for lower pulse rate applications it is usual to use a repetitively firing flash lamp combined with a lithium niobate Q-switch. Current performance at 20 pulses per second are about 1 Joule of energy in each pulse which lasts for 10 ns.

When very high energy pulses are required Nd–Glass rather than Nd–YAG is sometimes preferred. The very high gain, of about thirty, of the Nd–YAG laser means that when very high energy output pulses are required by Q-switching, energy within the cavity is lost by virtue of the spurious reflections within the resonator. The gain of the laser with glass as the host material is lower and so more energy can be built up inside the resonator before Q-switching. Furthermore glass is available with very high optical homogeneity thus reducing the possibility of optical and thermal damage. Because of the relatively ill-defined positions of the energy levels associated with the Nd^{3+} ions within the glass compared with YAG the upper laser level is wider and more energy can be stored although of course the linewidth of the output is wider and can be as much as 100 Å (10,000 GHz). Glass can also be doped to a much greater degree than the other two hosts, 6% compared with 1·5% for YAG and 2% for calcium tungstate.

Glass lasers are often used in conjunction with amplifiers and for outputs where only one or a few pulses per minute are required. Outputs from such combinations have reached 5000 J in a 3 ms pulse train. By Q-switching 1500 J in a 5 ns pulse and by mode locking 350 J in 20 ps can be achieved.

6.3. Semiconductor Lasers

A brief discussion of the basic principles underlying the conduction of electricity is necessary in order to explain the fundamentals of the semiconductor laser. The action of this laser is quite different from those which have been considered previously although the fundamental necessity for establishing a population inversion between two energy levels still holds.

An atom in free space can be represented by a set of sharp energy levels corresponding to possible energy states of the atom or the possible energy states of the electrons of each atom. When two atoms are in close proximity each set of energy levels is displaced slightly with respect to the other ; each level is said to have been split. If a very large number of atoms are in close proximity, as is the case in a solid such as a crystal, then in general, the energy levels associated with one free atom will split up into as many levels as there are atoms in the body. Figure 6.4 shows the situation for two energy levels in the free atom. Each energy level becomes a band of levels in the solid state.

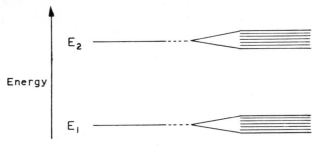

Fig. 6.4. The splitting of two atomic energy levels into bands in a crystal.

Conduction takes place by the movement of electrons under the action of an applied field. The field imparts energy to the electron which jumps from one allowed energy level within the band to another higher up. This, in turn, results in a vacant space or ' hole ' being left behind into which another electron from lower down can jump. This process continues and constitutes the passage of an electric current. For most solids, a large number of bands will exist with the electrons first occupying bands having the lowest energy. Eventually at a sufficiently high energy a band will exist which is either partially filled or completely filled. Such a band is called the valence band. In each case the next higher band will be empty and is called the conduction band. At room temperatures, of course, this is not strictly true because thermal energy will result in some electrons occupying the conduction band.

In the situation where the valence band is only partially filled, electrons can easily move up to the adjacent higher energy levels within the band and so support conduction. This is the situation with metals such as copper which are good conductors. If, however, the valence band is completely full, electrons will not be available for conduction unless they can reach the empty conduction band. Every material conducts to some extent because there is always a finite probability of thermal excitation to any higher level, but as the thermal energy at room temperature kT is only about 0·025 eV, the energy gap between the conduction and the valence band must be reasonably small. A class of materials called intrinsic semiconductors exists in which the separation between the conduction and valence band is about one electron volt. This results in some electrical conductivity being possible as a reasonable number of electrons are thermally excited to the conduction band (the number can be calculated by the Boltzmann distribution equation 1.4 given earlier). Materials other than conductors and semiconductors have valence bands separated from conduction bands by considerably larger energy differences—very few electrons are in the conduction band and hence these materials are insulators. The conductivity of semiconductors falls somewhere midway between that of conductors and insulators. This band theory of conduction accounts most satisfactorily for the observation that the conductivity of semiconductors increases with temperature (due to more electrons being thermally excited from the valence band into the conduction band) while the conductivity of ordinary conductors falls with increasing temperature (because increasing the temperature does not increase the number of electrons available for conduction, it only increases their probability of collision and hence the resistivity).

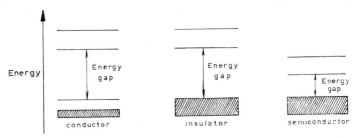

Fig. 6.5. Valence and conduction bands for three different types of materials. Shaded areas indicate levels occupied by electrons.

Figure 6.5 shows each of the cases described above. In order to avoid confusion, no electrons are shown in the conduction band of the semiconductor—this will only happen in reality at the absolute zero of temperature when a semiconductor will obviously be an insulator.

The semiconductor laser makes use of what is known as an extrinsic semiconductor. If an intrinsic semiconductor has atoms of some foreign material diffused into it by a process called doping, its energy diagram acquires new levels which lie between the conduction and valence bands. Semiconductors with these additional levels are said to be extrinsic. When a material, whose atoms have one valence electron more than the atoms forming the host lattice, is diffused into the host lattice, the result is called an n-type material. Spare electrons will therefore become available and will give rise to energy levels close to, but just below, the conduction band ; (in the case of silicon doped with phosphorus the energy gap is about 0·05 eV). Such an impurity is said to result in donor levels as it enables the material to act as a source of electrons available for conduction.

Figure 6.6 shows one such donor impurity level. If there are 10^{22} atoms per cubic centimetre and 0·1% doping is used then there will be 10^{19} such donor levels per cubic centimetre. The effects of such levels are considerable because thermal excitation will easily populate the conduction band from the donor levels enabling conduction to take place by means of transfer between energy levels in the conduction band. In a similar manner the host can be doped with a material whose atoms have one electron less than those of the host atom. This type of doping produces what are called p-type materials. With a

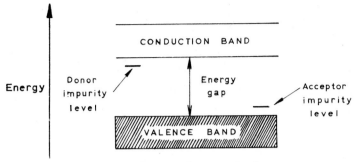

Fig. 6.6. Donor and acceptor levels.

p-type material positive holes are created whose energy levels lie close to, but just above, the valence band and are known as acceptor levels (in the case of silicon doped with aluminium the energy gap is about 0·08 eV). Electrons are then easily thermally excited from the valence band into the impurity levels and conduction can take place by hole movement within the valence band ; (see fig. 6.6).

A semiconductor laser is made by forming a junction between *p* and *n*-type materials in the same host lattice so as to form what is known as a *p-n* junction. The doping is extremely heavy, of the magnitude indicated above, so that the lower part of the conduction band of the *n*-type material is actually filled with electrons and the top part of the valence band of the *p*-type material is filled with holes. A voltage sufficiently high to overcome the energy gap *V* shown in fig. 6.7 is then applied to the junction so that the *n*-type region is connected to a negative supply and the *p*-type region to a positive supply. Under these conditions the junction is said to be forward biased. The electrons in the *n*-type region and the holes in the *p*-type region are then driven towards the junction where they combine to produce photons.

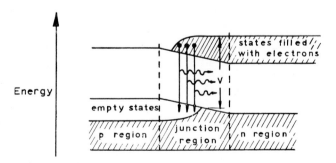

Fig. 6.7. Semiconductor laser—junction region.

Without the forward bias few photons would be produced as an electron would have to climb a potential barrier before it could re-

103

combine with a hole to form a photon. As the forward bias is increased more and more photons are emitted and so the light intensity becomes stronger.

The production of photons by applying electric fields to solid materials has been known for a considerable time, long before the invention of the laser, the phenomenon being known as electro-

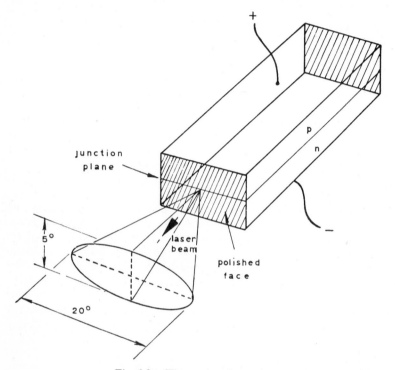

Fig. 6.8. The semiconductor laser.

luminescence although of course the emitted light is incoherent. Electroluminescence was known to be particularly strong in the material gallium arsenide and in 1962 several groups of workers reported laser action by passing extremely high currents through the junctions and simply polishing the end faces of the gallium arsenide p–n junction so that they acted as mirrors[68,69]. The sides of the device were roughened to prevent lasing in an unwanted direction and so wasting population inversion. Figure 6.8 shows the physical appearance of a gallium arsenide semiconductor laser although it must be borne in mind that the device is only a millimetre or so in size and the width of the emitting junction is only about $2\,\mu$m. Subsequently many other materials were found to be suitable for making p–n junction lasers, e.g. InP[70], InAs[71], InSb[72], PbTe[73], PbSe[74] and PbS[75].

As the current across the *p–n* junction is gradually increased from zero spontaneous emission (electroluminescence) starts and the output has a wide linewidth. As the current continues to be increased a time is reached when the spontaneous emission gives way to laser action and the linewidth of the output narrows dramatically. The lifetime of the electron-hole pair before recombination to emit a photon is extremely short, about 10^{-9} s–10^{-10} s, consequently the threshold currents required for the gallium arsenide laser are extremely high if sufficient population inversion is to be obtained. It is found that the threshold current density is constant at about 200 amps cm^{-2} below 20°K and rises rapidly as the temperature is increased. At the temperature of liquid nitrogen (77°K) the threshold is 750 amps cm^{-2} while at room temperature of 300°K as much as 50,000 amps cm^{-2} is required. Thus it can be seen that cooling the device produces more efficient operation as well as aiding the removal of dissipated heat. This latter consideration is extremely important and tends to set the limit on the power output of the device so in recent years considerable effort has been applied to heat-sink technology. A consequence of the short lifetime of the electron in the upper state mentioned above is that semiconductor laser output can be modulated easily by simply switching on and off the applied current. The output will faithfully follow modulation of the current up to frequencies of 10^9 or 10^{10} Hz[76]. This is an important advantage over other types of laser particularly for applications such as communications and high pulse repetition radar.

The energy gap V for gallium arsenide is 1·4 eV and so the corresponding wavelength emitted by the laser is at 8400 Å. For any particular device the gap can vary in width by as much as 50 Å, i.e. the output can have a width of 5 nm, so that although the 'mirrors' of the solid-state semiconductor laser are very close together compared with, say, a gas laser, for instance a 1mm separation results in an axial separation of 3 Å, many axial modes still oscillate. Linewidths of individual modes of less than 300 MHz have been reported[77]. The losses for transverse modes are very small and the output beam has considerable divergence as indicated in fig. 6.9.

Two efficiencies can be quoted for the gallium arsenide laser. First the percentage of photons obtained with respect to the number of electrons crossing the junction; this is called the external quantum efficiency and varies with temperature such that at room temperatures it is 15%, at 77°K it improves to 40% and at liquid helium temperatures it can be as high as 60%. A second way of defining efficiency, more useful for comparison with other types of laser is as a ratio of the laser power output to the electrical power input where efficiencies of as high as 10% have been claimed. In terms of actual power outputs, continuous and pulsed, the following figures will give an indication of the performance of the gallium arsenide laser. At 4°K a continuous power output of 15 W has been obtained[78] decreasing to a few watts at 20°K[79] and to one watt at 77°K[80]. At room temperatures only very low powers can be obtained continuously because of overheating[81]. By pulsing the laser a peak power of 100 W has been achieved at 77°K and 20 W at room temperatures.

A disadvantage of the gallium arsenide laser is that the output is invisible in the near infrared. For the penalty of higher threshold currents the wavelength of the output can be reduced to $0.64\,\mu$m in the visible red by doping the gallium arsenide with phosphorus and operating at liquid nitrogen temperatures[88].

So far discussion has been restricted to semiconductor lasers made of gallium arsenide for historical convenience. Recently, however, more complex lasers have been developed which have a higher performance. Gallium arsenide is still used but losses in power have been reduced. The losses arise chiefly from the narrow ($2\,\mu$m) emitting area causing laser light to be widely diverged into regions of the semiconductor where population inversion has not been sufficiently high for lasing. Consequently this laser radiation is absorbed and the output diminished. This problem has been alleviated by developing so-called heterostructure lasers in which a material of differing refractive index is placed close by and parallel to the plane of the junction—somewhat like a flat fibre optic. For obvious reasons this is known as a single heterostructure (SH) laser and threshold currents at room temperature are reduced to 10,000 amps cm^{-2}. In a single heterostructure device the added material is gallium aluminium arsenide deposited on the p-type side of the junction by vapour deposition or diffusion. By carrying out a similar operation on the n-type side as well then even more improvement is possible. Such double heterostructure (DH) devices as they are called have threshold currents as low as 1000 amps cm^{-2} at room temperature. Typical pulsed SH devices are 10% efficient giving 100 mW mean power at room temperatures increasing to 50% efficiency at 77°K while pulsed DH lasers can be as efficient as 25%. When continuous operation at room temperature is required DH lasers have the best performance, 120 mW at 7% efficiency having been reported. The great advances in microelectronics technology and the small size of semiconductor lasers has resulted in recent efforts to produce arrays of laser diodes. RCA in the United States have built arrays of 600 diodes giving $2\,\mu$s long peak pulses of 1.5 kW (30 W mean) with repetition rates of 10 kHz at 77°K.

In summary it can be said that the semiconductor laser is by far the most powerful laser by size and can be modulated easily at high frequencies. By choice of material a wide range of wavelengths is available, over two orders of magnitude from $0.33\,\mu$m to $34\,\mu$m and tuning is possible by varying either an applied magnetic field or the temperature or the pressure. Where high spatial and temporal coherence are required the semiconductor laser is not the best choice but for cheapness, compactness, ease of modulation and high efficiency it is supreme.

CHAPTER 7

gas lasers

Owing to their high efficiencies, wide choice of wavelengths, comparative independence of environmental conditions and outputs which are a close approach to an ideal coherent source of light, gas lasers are perhaps the most useful and certainly the most ubiquitous type of laser[90].

Gas lasers can be classified into three different types depending on the nature of the energy levels in between which laser action takes place. In the case of the helium-neon and the helium-cadmium laser a transition between unionized atomic states is involved, the argon and krypton lasers utilize a transition between ionized states and the carbon dioxide laser uses levels which arise from molecular rotation and vibration.

The two most significant mechanisms whereby an atom in a gas can be excited to a higher energy state are known as collisions of the first and second kind[91,92]. Collisions of the first kind involve the interaction of an energetic electron with an atom in the ground state. The impact of the electron causes it to exchange some of its energy with the atom and the latter becomes excited. This process can be represented by means of the following equation :

$$A + e_1 = A^* + e_2 \qquad (7.1)$$

where A signifies the atom in its ground or lower state and A^* in its excited state. The energy of the electron before and after the collision is designated by e_1 and e_2 respectively.

A collision of the second kind involves the collision of an excited atom in a metastable state with another atom of a different element in an unexcited state, the final result being a transfer of energy from the metastable state to the unexcited atom thus exciting it. The metastable atom therefore loses energy and reverts to a lower state. A collision of the second kind is represented by the following equation :

$$A_1 + A^*_2 = A^*_1 + A_2 \qquad (7.2)$$

where A_1 signifies the unexcited atom of an element labelled 1 and A_2 an unexcited atom of an element labelled 2, the asterisks indicating excited states.

7.1. The Helium-Neon Laser

The first gas laser to be constructed was a helium-neon laser built by Javan in 1960[16]. This type of laser is probably the most common and

108

so its operation will be described in some detail. It depends for its operation on a population inversion between two excited levels of the neon atom. An energy level diagram of the helium-neon system is shown in fig. 7.1. The helium energy levels are labelled using Russell-Saunders notation and the neon levels using Paschen notation. Space precludes an explanation of these here. The reader is advised to consult one of the standard text-books[93] although a knowledge of terminology is not essential for understanding laser action.

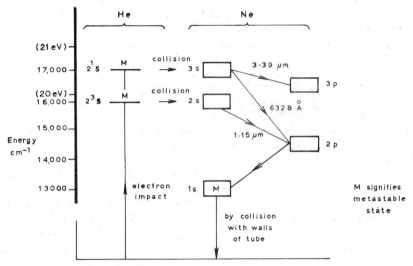

Fig. 7.1. Helium-neon laser energy level diagram.

The neon bands shown actually consist of many lines and so give rise to a laser output of many wavelengths of which only the three most powerful are indicated.

The helium atoms in the mixture of helium and neon are excited to the metastable states 2^1S and 2^3S by collisons of the first kind caused by an electrical discharge applied to the laser tube which is normally some 30 to 50 cm long. Using the nomenclature of equation 7.1 this process can be represented as follows ;

$$He + e_1 = He^* + e_2 \qquad (7.3)$$

These two energy levels are very close to the 2s and 3s levels of neon and collisions of the second kind then take place leaving the 2s and 3s levels well populated. This collision can be written as :

$$He^* + Ne = He + Ne^* \qquad (7.4)$$

The 2^3S level of helium is actually at 19·82 eV while the 2s level of neon is 19·78 eV. The 2^1S level of helium is at 20·61 eV and the 3s level of neon is at 20·66 eV.

109

Javan's first laser used the transitions 2s to 2p for laser action and a variety of wavelengths were obtained in the near infrared, the strongest line being at 1·1523 μm. Soon afterwards the familiar red line at 6328 Å which arises from a transition between the 3s and 2p states was discovered[94]. It was also observed, at about the same time, that a very strong output could be obtained further out in the infrared at 3·39 μm, the transition 3s to 3p being responsible[95]. All these transitions involve energy levels which are well above the ground state and so the helium-neon laser is a four-level laser.

It can be seen that the 3·39 μm transition shares a common upper level with the 6328 Å transition and so the 3·39 μm line will have an adverse effect on the power available in the visible line. Because of its much longer wavelength, the threshold of the 3·39 μm line is much lower and hence its gain is always much higher compared with the 6328 Å line. It is so high, in fact, that laser action can be obtained without the feedback provided by the laser mirrors being necessary, this effect being called superradiance.

In lasers required to operate with the maximum of power at 6328 Å, the 3·39 μm output must be suppressed[96]. Suppression can be carried out in two different ways, most commercial lasers employing either or both methods. One method is to insert a prism into the cavity and adjust its orientation for maximum output at 6328 Å thus effectively misaligning the laser for the 3·39 μm wavelength[97]. In some cases two prisms are used at each end of the cavity to prevent a double pass. A second method is to reduce the height of the 3·39 μm gain curve by increasing its width. This is done by placing a series of magnets along the side of the laser tube[98]. The inhomogenous magnetic field thus produced smears out, and so broadens, the 3·39 μm gain curve by virtue of the Zeeman effect. The 6328 Å is broadened as well but reference to equation 3.43 will show that the Doppler width of the latter is six times as broad as the 3·39 μm curve so the relative effect on the 6328 Å gain curve is very small.

Atoms in the terminal level 2p decay radiatively to the 1s metastable state in 10^{-8}s, much faster than the spontaneous rate of decay from the 2s to the 2p level which has a lifetime of about 10^{-7}s ; the lower lasing level is thus kept comparatively empty and so the conditions for population inversion with respect to the 2s level are satisfied.

It is inevitable that some population of the upper levels of neon will take place by direct excitation through collisions of the first kind. Thus the preferential filling of 3s and 2s levels which occurs by virtue of collisions of the second kind with the metastable helium atoms will be suppressed. Because of this, it is advantageous to increase the latter effect in comparison with the former by increasing the density of the helium atoms with respect to the neon atoms. This is achieved, in practice, by filling the laser tube with a pressure of 1 torr of helium and 0·1 torr of neon.

An additional important factor to consider is that the 1s state of neon is also metastable and so its population will tend to increase until, if it becomes too high, photons emitted by decay from the 2p level to the 1s level will have a high probability of exciting atoms in the 1s level back up to the 2p level again. This process is called radiation trapping and is similar to resonant radiation mentioned in Chapter 1. In this case the lifetime of the 2p levels is effectively increased to the detriment of the efficiency of laser action, since a reduced population inversion will result. The population of the metastable 1s state therefore has to be reduced by another process which, in order to avoid trapping between the ground and 1s levels, must be non-radiative. This is achieved by collision with the walls of the laser tube. It is for this reason that the gain of the helium-neon laser has been found to be inversely proportional to the diameter of the laser tube so that small tubes only a millimetre or so in diameter are used[99].

A diagram of a typical helium-neon laser is shown in fig. 7.2[100]. The Doppler width of the 6328 Å transition is 1700 MHz at room temperature so for ordinary tube lengths of tens of centimetres several axial modes will oscillate.

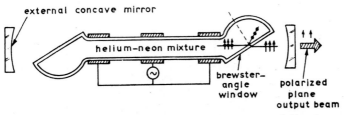

Fig. 7.2. The components of a gas laser (a helium-neon laser).

Excitation may be either by means of a radio frequency or a direct current discharge. For an r.f. discharge external electrodes are used and power at 30 MHz is supplied as readily available transmitters form convenient supplies. However, r.f. discharges are liable to cause interference with other radio communications and also the strong fields round the electrodes tend to drive the gases into the walls of the tube so that the pressure drops and eventually the tube must be refilled. Most commercial helium-neon lasers employ a d.c. discharge of between five and fifty milliamps from internal electrodes. The mirrors forming the resonating cavity of a helium-neon laser will be typically a 23-layer dielectric mirror of reflectivity 99·5% and a 9-layer dielectric mirror of transmission 1% which forms the output mirror—assuming, of course, that the laser is to work at 6328 Å, which is usually the case.

In comparison with most other lasers the helium-neon laser provides a relatively cheap source of high quality laser light but suffers from the drawback of small output. The practical upper limit of available

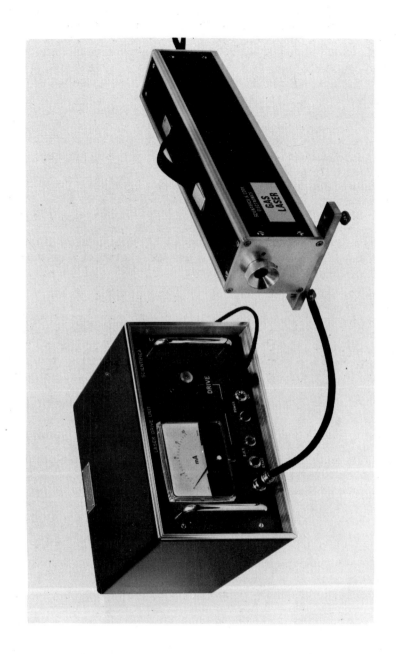

Fig. 7.3. A helium–neon laser with power supply giving an output of 5 mW uniphase. (Courtesy Scientifica and Cooke Ltd.)

power is roughly 100 milliwatts for each of the three major lines. A typical small helium-neon laser, as illustrated in fig. 7.3 together with its power supply, gives a uniphase output of about 5 milliwatts.

High powers and short pulses can be obtained by pulsing the input to obtain 100 W for a few microseconds[101] and shorter pulses of 1 ns duration may be obtained by mode locking or cavity dumping.

7.2. *The Helium-Cadmium Laser.*

The shortest wavelength continuous output available from a commercially made laser is provided by the helium-cadmium laser[102]. It has two outputs at 4416 Å, where 50 mW are obtained, and at 3250 Å where 5 mW are available. The wavelengths are selected by appropriate choice of output mirror reflectivity.

The laser tube contains a mixture of helium gas and vapour of the metal cadmium through which a d.c. discharge is passed. The method of operation is very similar to the helium-neon laser. A major problem in any metal vapour laser is the maintenance of the correct distribution of vapour in the tube necessary for optimum output power. Under normal conditions the ionized cadmium atoms, which provide the energy levels between which laser action takes place, drift towards the anode. This phenomenon is known as cataphoresis and tends to produce a gradient in the ion concentration so that at only one point along the length of the tube is the concentration optimized. If the tube is wide enough, diffusion will oppose this effect and prohibit the build up of large concentration gradients. However the gain of the helium-cadmium laser is inversely proportional to tube diameter so small diameter tubes are desirable as well as being more practical. A fairly

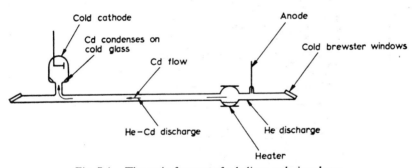

Fig. 7.4. The main features of a helium-cadmium laser.

uniform density of cadmium vapour is achieved by flowing cadmium down the tube towards the cathode thus counteracting cataphoresis. The gas tube itself is made hot enough by the discharge to prevent condensation of the cadmium vapour.

Figure 7.4 shows the essential components of a helium-cadmium laser. The laser tube is about 2 m long to obtain maximum gain. A heated

113

reservoir of cadmium at the anode end provides a source of ions which flow down the tube through the cathode where they condense on the cold walls on the other side of the cathode. About 3 watts per centimetre length of tube are dissipated so, unlike the argon laser which requires water cooling, simple convection cooling can be employed.

It has been found that if single isotope cadmium is used as opposed to a mixture of isotopes which make up the naturally occuring metal, higher outputs are obtained, particularly at 3250 Å, and a narrower gain curve of about 3 GHz is obtained.

7.3. *The Argon Laser*

Unlike the helium-neon laser the argon laser[103,104,105,106] works on a transition between two energy levels of the ionized atom. In order to singly ionize argon atoms, i.e. to remove one electron from each atom, a considerable amount of energy must be supplied to the argon gas. In consequence the power supplies for an argon laser are bulkier and more complex than for the helium-neon laser although the power outputs available are very much higher. However, the overall efficiency is about the same.

Figure 7.5 shows the energy level diagram of the argon ion laser and indicates two of the wavelengths in the visible which are available.

A much simplified idea of the mechanism by which atoms are brought into the appropriately excited state can be summarized by the following equation representing a collision of the first kind :

$$e_1 + Ar = (Ar^+)^* + e_2 + e_3 \qquad (7.5)$$

Depletion of the lower laser level is brought about by radiative decay to the ground state of the ionized atom followed by recombination with an electron to form the neutral atom.

Figure 7.8 shows the various lines available from the argon laser. The power depends on the current through the tube but about 80% of the total output power is approximately equally divided between the 4880 Å and 5145 Å. The Doppler widths of the argon lines are about 3500 MHz wide so for a given cavity length more axial modes will oscillate in comparison with a helium-neon laser.

Figure 7.6 shows the components of a typical argon laser[107] and fig. 7.7 a photograph of the working device. The discharge tube, made of beryllia, is 30 cm long and has a bore of 3 mm diameter. It is filled with argon gas at a pressure of 0·7 torr and subjected to a magnetic field of 1000 Gauss which constricts the discharge, increasing the electron density and so reducing the threshold. A stabilized current of about 40 amps at 165 volts is passed down the tube—a current density of some 500 A cm^{-2} being obtained. Such high current densities are necessary in order to ionize sufficient atoms, each of which require an excitation energy of 35·5 eV. A getter is incorporated to remove foreign gases. Because the device is d.c. operated positive ions tend to

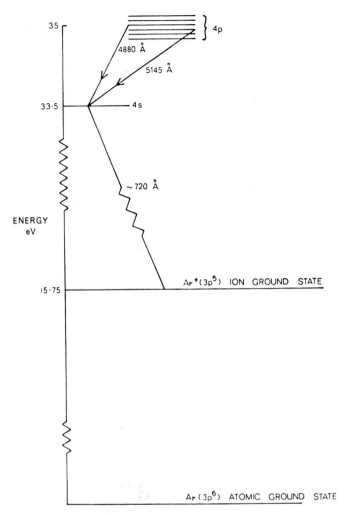

Fig. 7.5. The energy level diagram of the argon laser.

accumulate at the cathode end of the tube and, being heavier than electrons, the gas pressure tends to increase at that end ; a bypass tube is therefore provided to maintain an even pressure along the tube.

The laser described has a typical total power output of three watts of which approximately one watt is available in each of the lines 4880 Å and 5145 Å, the efficiency being 0·05%. Generally speaking all the lines are produced simultaneously with the usual mirror arrangement, although they all have widely different powers and thresholds, and so a prism cut at Brewster angles is incorporated as part of the cavity ; this

115

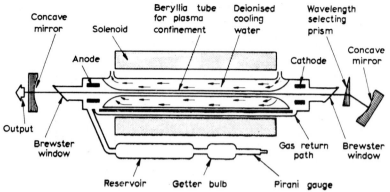

Fig. 7.6. Components of an argon ion laser.

makes the losses very high for all the wavelengths except one and the laser can be easily tuned to the desired wavelength by simply rotating the end mirror about a vertical axis.

The output power is greatly dependent on current density and considerable increases in power can be obtained by a relatively small increase in current. The latter is, however, limited by the rate at which heat can be dissipated, the anode and discharge tube are therefore usually cooled by the circulation of water. Materials, such as beryllia, which have a high thermal conductivity are therefore preferable to silica.

ARGON	KRYPTON
4370·73 Å	4577·20 Å
4545·04 Å	4619·17 Å
4579·36 Å	4680·45 Å
4657·95 Å	4762·44 Å
4726·89 Å	4765·71 Å
4764·88 Å	4825·18 Å
4879·86 Å	4846·66 Å
4965·10 Å	5208·32 Å
5017·17 Å	5308·68 Å
5145·33 Å	5681·92 Å
5287·00 Å	6471·00 Å
	6570·00 Å
	6764·57 Å
	6870·96 Å
	7993·00 Å

Fig. 7.8. Output wavelengths of argon and krypton lasers.

7.4. *The Krypton Laser*

This laser is almost identical to the argon laser except that the tube is filled with krypton gas[108,109]. This enables a new range of wavelengths

116

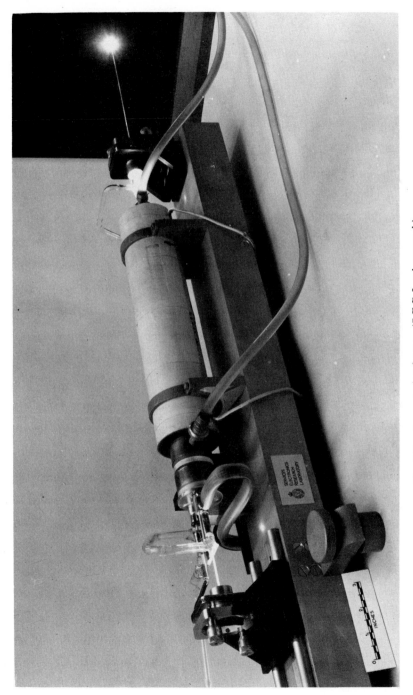

Fig. 7.7. An argon ion laser. (S.E.R.L. photograph).

117

to be obtained in the yellow and red as well as in the blue green. The output wavelengths available are listed in fig. 7.8 together with those obtainable from the argon laser[106].

By filling the tube with a mixture of argon and krypton an almost white output beam can be obtained.

7.5. *The Carbon Dioxide Laser*

This laser is by far the most efficient gas laser and the most powerful continuously operating laser. Efficiencies of 20% and outputs of several kilowatts are possible. The carbon dioxide laser works in the infrared at a wavelength of $10 \cdot 6 \ \mu m$[110,111].

In order to appreciate the theory of the carbon dioxide laser it is necessary to discuss the energy levels of the carbon dioxide molecule. It will be apparent that in order to obtain laser action in the infrared, energy levels whose separation is comparatively small must be found. Suitable levels are found in molecules which do not depend on electron excitation, but on the quantisation of the vibrational and rotational movements of the molecule. The carbon dioxide laser actually uses two additional gases : nitrogen and helium, the role of nitrogen[112] being similar to the role of helium in the helium-neon laser. The reason for using helium, as well, will be discussed later.

The carbon dioxide molecule[113] can be pictured as three atoms which usually lie on a straight line, the outer atoms being of oxygen with a carbon atom in between. There are three possible modes of vibration ; in each case the centre of gravity remains fixed :

(1) The oxygen atoms may oscillate at right angles to the straight line—this is called, for obvious reasons, the bending mode.

(2) Each oxygen atom can vibrate in opposition to the other along the straight line. This mode is called the symmetric mode.

(3) The two oxygen atoms may vibrate about the central carbon atom in such a way that they are each always moving in the same direction— this is the so-called asymmetric mode.

Figure 7.9 shows diagramatically the three modes, the carbon atom being represented in black.

Each possible quantum state is labelled as follows : for the symmetric mode by 100, 200, 300, etc. ; for the bending mode by 010, 020, 030, etc. ; and for the asymmetric mode by 001, 002, 003, etc. Combinations of all three modes are possible, e.g. 342 but they need not concern us.

In addition to these vibrational modes the molecules can rotate and therefore quantized rotational energies are possible ; a set of rotational levels is associated with each vibrational level, these are labelled in order of increasing energy by J values, each value being either 0 or a positive integer.

To make this nomenclature clear fig. 7.10 shows the sets of energy

118

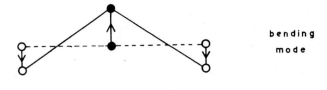

bending
mode

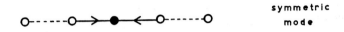

symmetric
mode

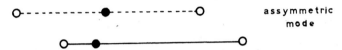

assymmetric
mode

Fig. 7.9.   Modes of oscillation of CO_2 molecule.

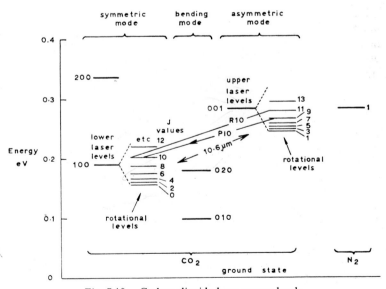

Fig. 7.10.   Carbon dioxide laser energy levels.

119

levels associated with each mode of vibration together with a set of rotational levels for the 001 and 100 modes on a much expanded scale. The ground state and the first excited state of the nitrogen molecule are also shown. As only two atoms are involved the nitrogen molecule can have only one vibrational mode.

The mechanism of laser action is as follows[114,115] : direct electronic excitation of the nitrogen molecule into its 1 state by a collision of the first kind. This process is represented as follows :

$$e_1 + N_2 = N_2{}^* + e_2 \tag{7.6}$$

A collision of the second kind with a carbon dioxide molecule in the ground state with excitation to the 001 state follows :

$$N_2{}^* + CO_2 = N_2 + CO_2{}^* \, (001) \tag{7.7}$$

This takes place because, as can be seen from the energy level diagram, the two energy values almost coincide. The 100 vibrational state is of much lower energy and so cannot be populated by this process.

The population of the 001 levels now exceeds the population of the 100 levels and so the population inversion condition for laser action to take place between these levels has been achieved. However, two points must be born in mind. First, a transition from the 001 level to the 100 level must obey a selection rule which states that J can only change by ± 1. Thus if $J = 10$ for a particular level then only the transitions from $J = 9$ to $J = 10$ and $J = 11$ to $J = 10$ are permitted. If J changes by $+1$ the transition is called a P-branch transition and if J changes by -1 it is called an R-branch transition. A transition from $J = 9$ to $J = 10$ is called $P10$ and a transition from $J = 11$ to $J = 10$ is called $R10$. Second, the population of the rotational levels of the 001 state will have a Boltzmann distribution, so, after taking degeneracy into account the effective population of a $J = 11$ level, for instance, will be less than the $J = 9$ level. The result of this is that P-branch transitions dominate because it so happens that a particular P-branch level will fill up (in order to restore equilibrium) by depletion of the population of the R-branch above it quicker than the R-branch level population decays by spontaneous emission to the lower laser level. The wavelengths associated with the most powerful transitions of the carbon dioxide laser at normal operating temperatures are : $P18 - 10\cdot57$ μm, $P20 - 10\cdot59$ μm, $P22 - 10\cdot61$ μm and the separation between each transition is about 55 GHz.

Each gain curve corresponding to a P-branch transition has a line-width of about 50 MHz. In comparison with other gas lasers this is a narrow Doppler width and comes about because the wavelength is some twenty times as long and the mass of the molecule is greater than that of most atoms. Reference to equation 3.43 will immediately show that these factors will reduce the Doppler width considerably. The sum of the areas under each gain curve in fig. 7.11 is proportional to the population inversion between the 001 and the 100 levels and hence proportional to the intensity of the output. These areas are not in fact equal and it so happens that because of the relative J-level populations the area under the $P20$ gain curve is largest. The axial mode separation for a 100 cm long cavity is by equation 4.21 about 150 MHz. Figure 7.11 shows the $P18$ and $P20$ gain curves and the axial mode spacing.

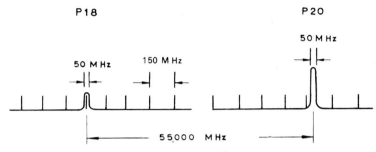

Fig. 7.11.   Gain curve of CO_2 laser output.

It is apparent from fig. 7.11 that where a tube one metre in length is used, only one axial mode can oscillate under a gain curve at any given time. If a much longer cavity were to be used the modes would be closer together and so several would oscillate. The mode which experiences the greatest gain will tend to grow in intensity at the expense of the others. This happens because the mode which starts to oscillate initially depletes the population of the appropriate 001 level and, as explained above, it so happens that the relaxation rate into such a depleted level from other J levels associated with the same vibrational level (in order to restore a Boltzmann distribution) is much faster than the spontaneous decay rate from any J level to a lower vibrational level. Hence the inversion between other levels tends to feed into the first. The gain profiles will uniformly decrease together and it follows therefore that the P-branch transitions are effectively homogenously broadened. For a short cavity where only one mode oscillates, the change in cavity length due to instabilities will cause the output power to fluctuate. If the laser is tuned so that the axial mode frequency is at the centre, for example, of the $P20$ gain curve, then a gradual reduction in power will be observed as the axial mode frequency drifts. If the next mode peaks up at $P18$ or $P22$ it will take over, so not only does the power fluctuate, but a frequency fluctuation is also obtained. On the other hand for the case of a 10 m cavity with a corresponding mode separation of 15 MHz, several modes will be present under each gain curve, and so the P-branch with a maximum gain always oscillates because one axial mode will always be present under the Doppler gain curve[116,117,118].

Inspection of equation 3.43 will show that by operating at a low

121

temperature the Doppler width of the gain curve is kept small and so the gain of the laser is increased[110,111]. The operating temperature can be reduced by restricting tube diameter to a few centimetres[121] and by adding helium to the mixture of carbon dioxide and nitrogen. The helium is effective in (a) increasing the thermal conduction to the walls of the tube, (b) indirectly depleting the population of the lower laser level 100 which is linked through resonant collisions with the 020 and 010 levels, the latter level being directly depleted by the helium and (c) by ' cooling ' the 001 rotational levels which results in the available population being more heavily distributed among the upper lasing levels. By water cooling of the tube and at the same time slowly circulating the gas along the tube a continuous output of 70 W per metre length is obtainable. Figure 7.12 shows an experimental carbon dioxide laser 15 m in length folded by means of mirrors halfway along the cavity so that the overall length is only 7·5 m. The tube diameter is 3 cm. Lasers of this size can provide 1000 W of 10·6 μm radiation continuously with output divergences of 5 mrads. Tens of kilowatts would seem to be the upper limit to lasers of this type. More powerful lasers can be made by increasing the flow of gas down the tube to several hundred metres per second so that tubes only a metre long can give outputs of several kilowatts. Even more powerful continuously operating carbon dioxide lasers can be constructed by employing transverse gas flow although this can give rise to problems of transverse mode quality.

An example of a commercially available carbon dioxide laser is shown in fig. 7.14. Here the cavity length of 14 m has been folded no less than twelve times so that the entire laser is contained in a box only 1·5 m long. 400 watts of continuous output is produced in the uniphase. TEM_{oo} mode with a beam divergence of only 2·5 mrads. This makes the instrument ideal for cutting and welding.

The most powerful continuously operating carbon dioxide lasers have been made by effectively supplying only excited molecules to the laser cavity. In the so-called gas dynamic laser, carbon monoxide or C_2N_2 gas is burnt with air and expanded supersonically through a nozzle which cools the gas more quickly than the excited state molecules can relax and thus establishes a population inversion. Gas dynamic carbon dioxide lasers have reportedly produced 60 kW multimode and 30 kW single transverse mode although efficiencies are only a few per cent.

Some interesting techniques have been developed for producing high power pulses. Five methods may be mentioned:

(i) The electrical discharge in the laser can be pulsed. This is the simplest most obvious method but is not used very often as power output is low and pulse time is long.

(ii) Q-switching by means of a rotating mirror (see fig. 5.7) or

Fig. 7.12. A 450 W c.w. carbon dioxide laser. (S.E.R.L. photograph).

saturable dye. Typically 30 kW peak pulse power per metre of discharge repeated at 5000 pulses per second each pulse lasting 150 ns.

(iii) Mode locking techniques can be used to produce pulses of duration less than a nanosecond.

(iv) Greater output can be obtained by increasing efficiency or by increasing the output from each unit volume of gas. As early carbon dioxide lasers were essentially low pressure devices and efficiencies were already quite high, attention was turned to increasing the pressure of the gas. At atmospheric pressures, however, it is most difficult to get a homogenous stable discharge. However, eventually a method was found whereby the electrodes, instead of being two, one at each end of the tube, consisted of comb-like structures along the side of the tube to produce a transverse discharge. In this way a very rapid discharge could be obtained which was uniform for some tens of nanoseconds before arcing at isolated points occurred. Such ' transverse excited atmospheric ' or TEA lasers produce very short pulses of great power, typically 100 MW for 50 to 100 ns at low repetition rates with 10% efficiency. Output pulse energies are greater than 100 J. TEA lasers are likely to have widespread use as the potential applications of carbon dioxide lasers are realized.

(v) Gas pressures above one atmosphere can be used by pre-ionizing the laser gas. This is done by directing a pulsed beam of electrons accelerated to a potential of several hundred kilovolts into the gas. The low energy electrons are dispersed uniformly throughout the gas and uniform discharges result. Pulses are of longer duration than TEA lasers, typically 20 μs and 100 MW power, i.e. 2000 J of energy per pulse.

The long wavelength at which the carbon dioxide laser works brings materials problems in that glasses suitable for use in the visible are often opaque at 10·6 μm, consequently materials for output mirrors, high reflection mirrors (i.e. nominally 100% reflecting) and optical components such as lenses and prisms must be chosen with care[127,128].

The output mirror is often made of germanium and has a transmission depending on the length of the tube—for a one metre tube a transmission of about 20% is usual. The germanium is coated on one side with a multilayer dielectric mirror and on the other with an antireflection coating to suppress undesirable reflections between the front and back surfaces. For high power work the absorption of germanium is too high, so mirror substrates made of alkali halides are used but suffer from a tendency to absorb water vapour. Ideally output mirrors should be made of materials which are low absorbing, readily polished to be flat to a wavelength or so and are suitable for antireflection and dielectric coatings. The various materials used are contrasted in fig. 7.13.

Material	Power Tolerance	Absorption (per cm thickness)	Reflectivity from one surface	Remarks
Germanium	> 200 W cm^{-2} (edge cooled)	~ 3% (at < 40°C)	36%	Good chemical stability. Insoluble in water.
Gallium Arsenide	> 1 kW cm^{-2}	2%	27%	Good chemical stability. Insoluble in water. Expensive.
Alkali Halides (eg. potassium bromide, sodium chloride)	> 1 kW cm^{-2}	< 0·01%	4%	Hygroscopic. Cheap. Difficult to coat. Easily polished.
Irtran (e.g. magnesium fluoride, zinc sulphide)	~ 20 W cm^{-2}	10%	17%	Insoluble in water. Rugged. Easily polished.

Fig. 7.13. Carbon dioxide laser output mirror materials.

High reflection mirrors must be deposited on stable substrates in view of the high power densities involved. High modulus of rigidity, low thermal expansion and good thermal conductivity are essential qualities. The substrate surface must be capable of being highly polished and suitable for evaporation. Stainless steel coated with gold (reflectivity 99%) or silica coated with aluminium (reflectivity 98%) are commonly used, although copper, sapphire and germanium are sometimes employed.

Lenses and optical components are usually made of germanium (refractive index 4·0) or sodium chloride (refractive index 1·5). Brewster windows may be made of sodium chloride or the mirrors can be attached directly to the tube by flexible metal bellows.

Fig. 7.14. A folded reconator carbon dioxide laser (by courtesy of Ferranti Ltd).

CHAPTER 8
the laser as a spectral source

8.1. *Standard of Length*

UNTIL 1960 the international standard of length was a metre bar made of platinum-iridium and kept at Sèvres in France. This length was based on the original definition of the metre which was defined in 1791 to be 10^{-7} of the quadrant of the earth's meridian passing through Barcelona and Dunkirk.

In 1960, however, the need for increased accuracy led to the International Conference on Weights and Measures to agree on a new standard which was defined at 1,650,763·73 times the wavelength of the orange line at 6057 Å emitted as a result of the transition $2p^{10}$ to $5d^5$ in isotopic krypton of atomic weight 86. With this standard a reproducibility of 1 part in 10^8 is possible but the low intensity and instability of the source place a limit on the reproducibility of measurements.

The advent of the helium-neon laser provided an intense spectral line of extremely small width. Nevertheless the essential property of any standard is reproducibility and the absolute linewidth of the laser depends very much on operating conditions[129,130].

One solution to this problem is to enclose within the cavity of a helium-neon laser a material which has a very sharp and well defined saturated absorption line within the Doppler width of the laser transition. The laser is thus constrained to operate at this line.

One of the most successful attempts to date[131] to achieve a stabilized laser output suitable for a standard was by Barger and Hall at the National Bureau of Standards. By placing a cell containing the gas methane in the cavity of a helium-neon laser and by careful servo control, they succeeded in reproducing the 3·39 μm output to an accuracy of 1 part in 10^{11}, i.e. at least a thousand times more reproducible than the Kr^{86} standard. The linewidth obtained was actually 160 kHz or 1 part in 10^9.

There is a distinct possibility therefore that in the near future standards of length will be redefined in terms of the stabilized output from a laser.

8.2. *Raman Spectroscopy*

Raman scattering was first observed in 1928[132] and has provided a powerful method for studying the structure of molecules. Until recently only incoherent sources such as mercury lamps were available but the increasing use of lasers has proved to be a major advance in the techniques of Raman spectroscopy[133,134,135].

Before discussing the advantages of lasers for Raman spectroscopy it will be useful to give a brief account of the Raman effect.

The energy levels associated with many molecules can be considered as essentially a ground level and an excited energy level corresponding to an electronic transition. In addition, there are a very large number of levels associated with the vibrational and rotational energies of the molecule. The energies of these levels are very close to each other in comparison with the separation of the electronic energy levels. As a result, the absorption spectra of such molecules consist of lines in the far infrared caused by excitation of vibrational or rotational states and in the ultraviolet corresponding to electronic transitions. Thus materials such as nitrogen, oxygen, and carbon tetrachloride are transparent in the visible and absorb strongly in parts of the ultraviolet and infrared.

To gain knowledge of the rotational and vibrational properties of such molecules, and hence also of their structure, a study of the absorption by a broad infrared source would normally be required. Such a technique would be very inconvenient compared with detecting absorption spectra in the visible part of the spectrum.

In 1928 C.V.Raman noticed that when visible light was passed through some materials such as those just mentioned, and the material was observed at right angles to the direction of propagation of the incident light, then scattered light of various different wavelengths could be seen.

Although molecules absorb strongly in certain discrete regions of the spectrum, incident light of any wavelength will always be scattered to some extent. This effect is known as Rayleigh scattering and can be understood by considering the molecule as a dipole which is forced into oscillation by the incident electromagnetic radiation. As an oscillating dipole emits radiation at the frequency of its oscillation, light will be re-emitted at the incident wavelength. Alternatively the electron can be envisaged as being excited from a lower state to a higher state and then dropping down to the lower state emitting light of the same wavelength uniformly in all directions.

Raman observed this Rayleigh scattering as a spectral line whose wavelength coincided with that of the incident light. However, very close to this line, other lines were also observed of longer and shorter wavelength. It was concluded that these lines arose as a result of the incident light quanta, of energy $h\nu_0$, interacting with the molecules in such a way that the rotational or vibrational energies were added to, or subtracted from, the energy of the incident photon. Therefore if the difference between two vibrational or rotational energy states is ΔE, the scattered photons have energies given by

$$h\nu = h\nu_0 \pm \Delta E \qquad (8.1)$$

Thus the frequency difference between the Raman lines and the source line is constant and is independent of the latter.

If the particular pair of vibrational or rotational states considered is entirely populated in the lower level then the minus sign applies and conversely for the positive sign. Unless the temperature of the material is very high then, by the Boltzmann distribution law, the lower energy levels will have much higher populations so photons of energy $hv = hv_0 - \Delta E$ will be more numerous and the lines of lower frequency than v will be brightest. These are, in fact, called Stokes lines in accordance with Stokes law of fluorescence which states that the fluorescent radiation is always of lower frequency than the incident radiation. Lines of shorter wavelength which exist as a result of molecules occupying the upper energy levels appear to contradict Stokes law and are therefore referred to as anti-Stokes lines.

The Stokes and anti-Stokes lines are, of course, always separated by the same frequency from the frequency of the incident radiation.

Figure 8.1 shows the energy levels concerned with the formation of Stokes and anti-Stokes lines.

	STOKES	ANTI-STOKES
Before interaction	$E = hv_0$ ΔE	$E = hv_0$ ΔE
After interaction	$E - \Delta E = hv$	$E + \Delta E = hv$

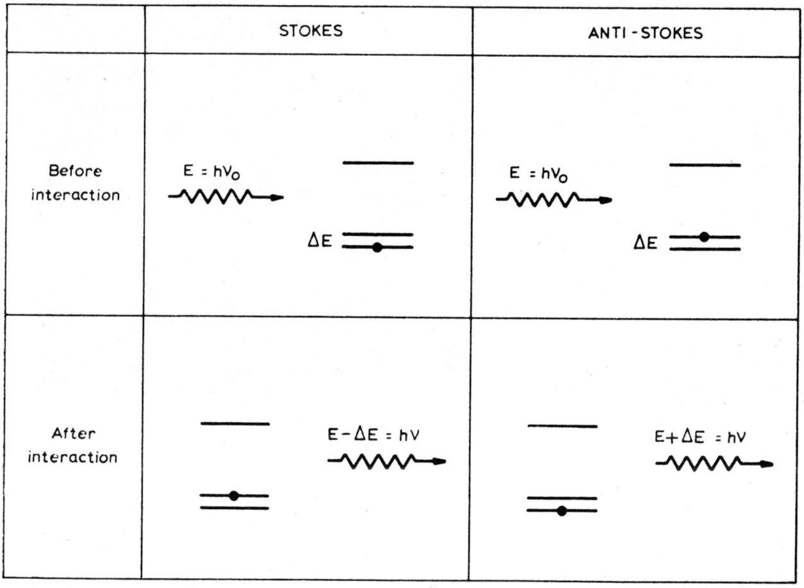

Fig. 8.1. Interactions leading to formation of Stokes anti-Stokes lines in Raman spectra.

It can be seen therefore that, by the Raman effect, the effects of vibration and rotation of molecules can be transferred from the infrared to the visible region of the spectrum thus facilitating examination and analysis.

The laser has the following advantages over conventional light sources for Raman spectroscopy :

(a) The Raman scattered light is only a small percentage of the incident light, consequently the very high intensities available from lasers enable lines to be observed which, if an incoherent source were used, would be far too weak.

(b) The large linewidths from intense conventional sources often render Raman lines, which are very close, to be unresolved. Here the extremely narrow linewidth of a laser greatly increases the resolving power.

(c) The Raman scattering increases as the frequency of the light source is increased and so powerful argon ion lasers with an output at the blue end of the spectrum are almost ideal sources.

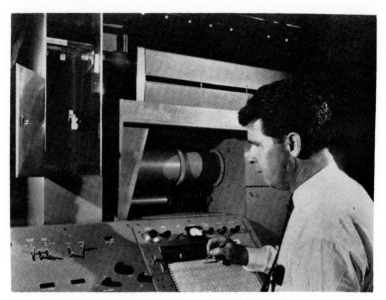

Fig. 8.2.   A 50mW helium-neon laser used as the light source in a commercial Raman spectrometer.   (Photograph by courtesy of Spectra-Physics, Ltd.)

(d) Raman spectra are usually recorded nowadays with a photomultiplier.   As these are most sensitive to blue wavelengths the argon laser is again a most attractive source.

(e) By studying the polarization of the Raman scattered light with respect to that of the incident light, the symmetry of the vibrational states of molecules can be ascertained.   For example if the scattered light is not polarized in the same direction as the incident light the vibrational molecular state is assymetric.   As the light from gas lasers is inevitably polarized information of this type is readily available.

(f) Fluorescence occurs when the excited atom returns to its ground state.   It may be that it returns via intermediate levels and the spectral

130

lines so produced can be confused with the Stokes lines of the Raman effect. By changing the type of laser used a different source wavelength can be used in a spectral region where such confusing effects do not occur.

(g) Although the power per unit frequency interval of the laser is very high the total power is much less than that from a conventional mercury lamp and so, with a laser source, photodecomposition is less likely.

Figure 8.2 shows a helium-neon laser being used in a Raman spectrometer.

8.3. *Stimulated Raman Scattering*

In 1964 Woodbury and Ng of Hughes Aircraft[136] were working with a ruby laser which was Q-switched by a Kerr cell containing nitrobenzene. They noticed that 10% of the output consisted of 7660 Å radiation, while the remainder was the usual 6943 Å output. The 7660 Å output was the result of interaction with the vibrational energy states of the nitrobenzene and was of the same wavelength as obtained by the conventional Raman effect. In this case, however, the Raman radiation was coherent and very much more intense than that obtained in the usual incoherent fashion. This process is known as the stimulated Raman effect and conversion efficiencies of up to 50% have been obtained.

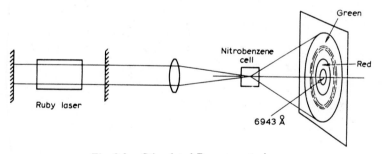

Fig. 8.3. Stimulated Raman scattering.

If the output from a Q-switched ruby laser is focussed onto a nitrobenzene cell placed externally, as shown in fig. 8.3, stimulated Raman emission takes place within the nitrobenzene and the gain is so high that powerful beams of light of different colours are obtained. Both Stokes and anti-Stokes lines are produced but the latter are obtained by obeying equation 8.2,

$$h\nu = h\nu_0 + k\Delta\nu \qquad (8.2)$$

where $k = 1, 2, 3$, etc. and $\Delta\nu = h\Delta E$.

The anti-Stokes lines are visible to the eye as red, yellow and green concentric rings as indicated in fig. 8.3. The concentric rings arise as a

131

result of the necessity for phase matching which was discussed earlier in Chapter 5 for the case of frequency doubling.

Stimulated Raman scattering has been observed in many materials and has provided coherent radiation of many new wavelengths.

8.4. *Tunable Lasers*

In this section we shall discuss lasers which can be continuously tuned over a range of wavelengths. Semiconductor lasers can be tuned by varying the position of the energy levels between which the laser action occurs. This can be achieved by varying the temperature or pressure of the laser material. Gallium arsenide lasers have been tuned over a small range by varying their temperature; lead selenide lasers, working in the far infrared, have been tuned from 7·5 μm to 22 μm by varying the pressure on the semiconductor between 1 and 14000 atmospheres[137]. However the most promising variable wavelength lasers are those based on parametric amplification and on employing liquids as the active medium. The practical possibility of such lasers is important because there are optical instruments whose performance would benefit enormously by replacing the conventional light source with a tunable laser. Absorption spectrometers, for example, employ a white light source and narrow regions of the spectrum are selected by a narrow band filter which usually takes the form of a diffraction grating. The amount of light obtained at any one wavelength is clearly only a very small fraction of the total light energy emitted by the lamp. The use of a laser would enable all the energy to be concentrated into any required region of the spectrum. Further the narrow linewidth of the laser would result in much higher resolution. Flash photolysis, in which fast chemical reactions and short lived chemical compounds can be examined, would also clearly benefit from a pulsed tunable laser.

8.4.1. *Parametric Oscillators*

Frequency doubling using non-linear optical effects[138] has already been described in Chapter 5 and is a particular example of what is known as a parametric effect. Consider two sinusoidally varying electromagnetic fields of different frequencies denoted by $E\omega_1$ and $E\omega_2$ and with frequencies ω_1 and ω_2 respectively, then the polarization produced by these fields acting together can be expressed in the form :

$$P = \chi E\omega_1 E\omega_2 \tag{8.3}$$

where χ is the susceptibility. It can easily be deduced that electromagnetic waves at two new frequencies ω_3, ω_3' will be produced and these frequencies are given by

$$\omega_3 = \omega_1 + \omega_2 \tag{8.4}$$

$$\omega_3' = \omega_1 \sim \omega_2 \tag{8.5}$$

Such effects only occur in non-centrosymmetric materials[49].

In the case of frequency doubling $E\omega_1$ and $E\omega_2$ are identical and

132

equations 8.4 and 8.5 hold to give a new wave of frequency 2ω and a d.c. component respectively.

$E\omega_1$ and $E\omega_2$ are often known as the pump and the signal—the pump being of higher frequency.

The first demonstration of optical parametric amplification was achieved by Wang and Racette[139]. The pump consisted of a 2 MW pulse at 3472 Å obtained from a ruby laser by second harmonic generation in ADP. The signal was a 10 mW beam from a helium-neon laser and the pump and signal were mixed collinearly in an 8 cm long ADP crystal as shown in fig. 8.4.

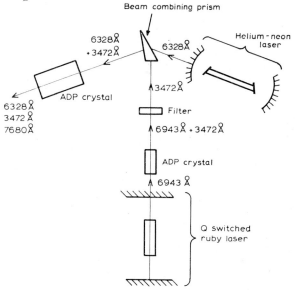

Fig. 8.4. Parametric amplification.

The emergent beam consisted of radiation of three different frequencies : at the pump and signal frequencies and at a third frequency, called the idler, corresponding to the difference in frequencies between the pump and the signal. In this experiment the idler consisted of an output of 1·2 mW at 7680 Å.

It can be seen from equations 8.4 and 8.5 that if ω_1 and ω_2 are fixed then ω_3 is also fixed as a sum or difference frequency. However, if only one of the three frequencies is fixed then the other two are free to range over many values, provided the sum of their frequencies is equal to that of the fixed frequency. In practice the two variable frequencies are decided by the particular phase matching used. Only one pair of frequencies can be phase matched at a time. By adjusting the phase matching parameters, e.g. the temperature of the non-linear crystal, the laser can be tuned over a range of frequencies.

Parametric oscillation was achieved first at the Bell Telephone Laboratories in 1965 when Miller and Giordmaine[140] used a neodymium doped calcium tungstate laser giving 10 kilowatts of power in the form of a pulse at a wavelength of 0·53 μm achieved by frequency doubling. A 5 mm long lithium niobate crystal was used as the parametric conversion medium. The output was tuned by changing the temperature of the crystal, an 11°C change in temperature produced output frequencies ranging from 9700 Å to 11500 Å. The optical arrangement used is shown in fig. 8.5. In Russia, at about the same time, ADP and KDP were also used as parametric conversion media.

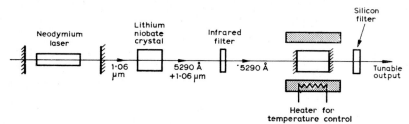

Fig. 8.5. Arrangement of components in first optical parametric oscillator.

Owing to the poor conversion efficiencies, a pulsed pump was needed in the above cases. Continuous parametric amplification was achieved **at Bell Telephone Laboratories**[141] **using barium sodium niobate** in place of lithium niobate, the latter also suffering from optical damage. Figure 8.6 shows the various components of the system.

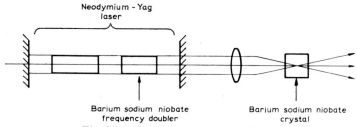

Fig. 8.6. The c.w. parametric oscillator.

A neodymium—YAG laser was used and included a barium sodium niobate frequency doubler within the cavity. The 0·53 μm radiation produced was focussed by a lens onto a second barium sodium niobate crystal. The temperature of the latter was varied from 97° to 103°C to produce a shift in wavelength from 9800 Å to 11600 Å corresponding to a change in bandwidth of $4·5 \times 10^{13}$Hz. It was found that only 45 mW of pump power at 5300 Å were required to produce parametric conversion and that 300 mW of pump power gave an output of 3 mW so that the efficiency of the process was 1%. More recently[142,143] parametric oscillators have achieved efficiencies of 50% pulsed and 30% c.w.

134

It seems probable that parametric devices will produce tunable coherent light[144] of wavelength down to the ultraviolet and that crystals which do not absorb in the infrared, such as proustite[145] Ag_3AsS_3, will enable the other end of the spectrum to be similarly exploited.

8.4.2. Liquid Lasers

Some liquids which are usually organic dyes can be made to lase. The use of liquids as the active medium of a laser is obviously attractive insofar as liquids can be of very high optical quality and the cooling problems which often arise in solid-state media can be overcome by circulating the liquid. Furthermore it has been found that liquid lasers can be tuned easily by a number of different methods over a wide range of frequencies. High quality solid state laser crystals are usually extremely expensive whereas the cost of suitable liquids is almost negligible. Although none of the lasers described in this section is available commercially, liquid lasers have a large potential and so an account of their development is merited.

Attempts to make liquids lase were made early on in the history of laser development but were largely unsuccessful because the heat from the pumping source caused inhomogenous heating of the liquid. The associated variations in refractive index resulted in very high scattering losses.

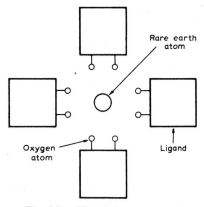

Fig. 8.7. The chelate structure.

Many organic molecules have broad, efficient absorption bands and so facilitate the establishment of a population inversion by pumping. Unfortunately a liquid laser cannot be made by dissolving the organic material in any solvent as the vibrations of the molecules making up the liquid usually rapidly depopulate the upper laser level and so quench laser action. One method of stopping this is to combine a rare earth atom such as europium with a chelate molecule. Figure 8.7 shows a schematic diagram of a chelate molecule.

The rare earth atom is in the centre of the structure surrounded by four pairs of oxygen atoms, these have the effect of binding the rare earth atom to the four organic structures which are known as ligands. This structure effectively isolates the rare earth atom from the liquid molecules.

The first successful liquid laser was reported in 1963 by Limpicki and Samelson[146]. They used the rare earth europium and benzoylacetonate as the ligand. The complete molecule was dissolved in alcohol to produce a thick viscous liquid. A xenon flash tube was used to pump the laser, a minimum pumping energy of about 2000 J being required. By operating the laser at between $-163°C$ and $-133°C$ a red output at 6131 Å was obtained.

A simplified energy level system for a chelate molecule is shown in fig. 8.8.

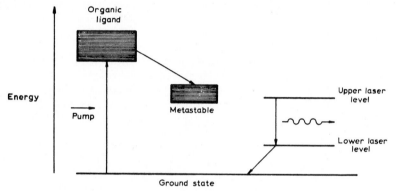

Fig. 8.8. Energy level system of the chelate molecule.

The organic ligand strongly absorbs pumping radiation. The excited molecules produced can either drop directly to the ground state or, in the case of molecules containing rare earth ions, most of the excited molecules relax to an intermediate metastable state which almost coincides in energy with the upper lasing level of the rare earth ion. The consequential exchange of energy between these levels pumps the laser.

The rapid temperature changes of the laser liquid make for difficulties in mirror alignment. This can be overcome by using the arrangement shown in fig. 8.9.

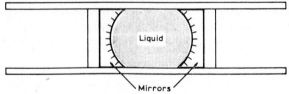

Fig. 8.9. Piston arrangement for the compensation of thermal expansion of the liquid in a liquid laser.

The tube containing the laser liquid is fitted with two quartz pistons upon which the mirrors are mounted. As the liquid expands and contracts the pistons are drawn up and down the tube while maintaining accurate alignment.

By using a different ligand, benzoyltrifluoroacetonate, dissolved in acetonitrile solution, operation at room temperature has been achieved. Nevertheless interest in chelate lasers has waned as it was found that the chelate structure absorbed the pump radiation to such an extent that only the surface of the liquid could achieve population inversion and the resulting low gain restricted output powers.

The next advance in liquid lasers was to use a solid state ruby laser to provide an intense pumping source. Sorokin and Lankard[147], in 1966, used this method to pump an organic substance called chloro-aluminium phthalocyanine which was dissolved in ethyl alcohol. Coherent infrared radiation was emitted at 0·755 μm.

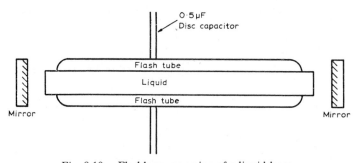

Fig. 8.10. Flashlamp pumping of a liquid laser.

Since then many other organic liquids have been discovered[148] which, when pumped by a laser, emit coherent light. Probably the most successful has been a substance called Rhodamine 6G which has provided the highest energy outputs and efficiency so far of any liquid laser[149].

In 1967 a further advance was made when Sorokin and Lankard[150] used an incoherent flashlamp to pump a non-chelate liquid laser. Essential conditions for a liquid laser pump are short pulse duration, high energy and fast risetime. Although the risetime of the best flashlamps is some 10^{-7}s compared with 10^{-9}s for a Q-switched ruby laser, Sorokin obtained laser operation in Rhodamine 6G dissolved in ethyl alcohol. The laser is depicted in fig. 8.10. It consists essentially of two coaxial quartz tubes, the outside diameter being 22 mm and the cavity being 3 mm across. The inner tube contains the liquid and the pumping flash occurs in the outer tube which, of course, completely surrounds the liquid. To obtain the shortest possible risetimes a very low inductance disc capacitor was used which straddled the tube and produced a flash of 100 J energy lasting 800 ns with a risetime of 300 ns. The output energy was 70 mJ in the yellow at 5850 Å and lasted for a few hundred nanoseconds. By flowing the liquid down the tube, repetition rates can be as high as one pulse per second.

The liquid lasers described so far can be tuned by varying any of several parameters. Changing the Q of the cavity, the solvent, the optical path length in the cavity or the pump energy can change the wavelength of the output. However, the widest tuning ranges are obtained by varying the concentration of the liquid.

The pumping light can be introduced down the axis of the laser when it is said to be longitudinally pumped or it can be pumped at right angles to the axis when it is said to be transversely pumped. Longitudinal pumping can be extremely symmetrical and so produce beams of narrow divergence, typically less than 0·5 milliradian. Further, by pumping longitudinally with a laser, efficiencies as high as 50% can be reached[151].

A narrow spectral line tunable over several hundred angströms may be obtained by using a diffraction grating as one mirror of the laser[152,153]. The first order beam from the grating is returned down the laser to the other mirror. As the direction of the first order diffracted beam from the grating depends on the wavelength, the latter may be tuned by tilting the grating as shown in fig. 8.11.

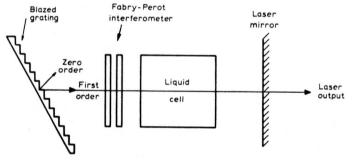

Fig. 8.11. Liquid laser using a grating mirror and Fabry-Perot interferometer to obtain an output of narrow linewidth.

The linewidths produced with normal mirrors are usually between 50 Å and 100 Å, the grating reflector reduces the linewidth to about 1 Å and by incorporating a Fabry-Perot interferometer within the cavity (included in fig. 8.11) the linewidth can be further reduced to about 10^{-2} Å.

One other important liquid laser remains to be described which uses an inorganic molecule as the lasing material. It was stated earlier that the vibration of the liquid molecules tends to depopulate the upper energy level and so prevent laser action. Although this is true, it is found that when a rare earth is dissolved in a liquid the rare earth atoms can become surrounded by a region known as a solvation shell which does offer some protection against external vibrations. It is also found that the efficiency with which photons are produced by transitions from the upper energy level, as compared to transitions imparting vibrational energy, is directly proportional to the square root of the mass of the lightest atom of the solvent. As most solvents contain hydrogen, few photons are normally produced. By using neodymium in the solvent selenium oxychloride, Lempicki and Heller[154,155] obtained laser operation (oxygen of mass 16 being the lightest atom in the solvent). The neodymium atom is more readily absorbed by the selenium oxychloride if antimony trichloride or tetrachloride is added. The laser produces an output at 10550 Å and is directly comparable with neodymium—glass lasers in terms of efficiency, power, ease of pumping and spectral width.

Continuous operation has recently been reported[156] in an aqueous solution of Rhodamine 6G flowed at 200 cm s^{-1} and excited by the 960 mW output from an argon laser. An output of 55 mW at 5965 Å with a half-width of 30 Å was obtained.

The development of liquid lasers is still in its infancy but the results so far show that high power continuous outputs tunable over a wide range of frequencies may be a reasonable expectation in the future.

CHAPTER 9

measurement with a laser

9.1. *Alignment*

WHEN a laser is operating in the uniphase transverse mode the output beam can be regarded as a straight line of almost constant thickness. This property can be utilized for the purposes of alignment. An accurate and easily repeatable method of alignment is necessary in construction work and engineering. Examples in which a laser has been used successfully include tunnel boring, pipe laying and bridge construction. Not only can a straight line be defined but qualities such as squareness and parallelism can also be ascertained.

For straight-edge applications the laser can be used in its own right or as a light source in conjunction with optical elements. Figure 9.1 is a photograph of a commercial helium-neon laser transit telescope which can provide a 10 cm diameter beam at a distance of 1·6 km or, in conjunction with a cylindrical lens, a line 10 cm wide by 250 m high at the same distance.

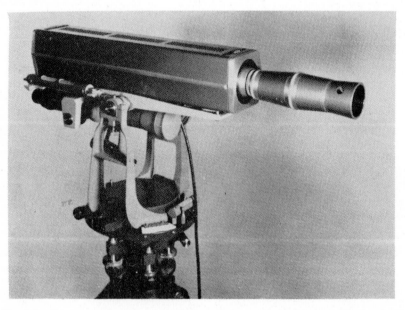

Fig. 9.1.   Laser transit telescope.   (Photograph by courtesy of Spectra-Physics, Ltd.)

140

It is obvious that the accuracy obtainable must depend on the ability to define the centre of the laser beam and for a readily definable centre, the symmetrical uniphase mode is obviously the most suitable as well as having minimum divergence. Thus the laser used must not only be sufficiently rugged to withstand the rigours of the environment in which it is to be used, but it must also provide a stable uniphase output. Stability is an important factor ; once the instrument is set up it must be capable of providing a beam which continues to point in the initially determined direction. The successful application of the laser in construction work may well depend on this feature as conventional theodolites are capable of high accuracy. Continuous checking of the work as it proceeds is easily undertaken with a permanently set up laser and relatively unskilled personnel, whereas the constant use of a theodolite is prohibitively expensive.

For applications where extremely high accuracy is unnecessary, such as tunnel boring, a crude method of estimating the beam centre can be used. This can take the form of a graticule of concentric rings placed in the path of the beam. The machine operator merely has to guide the machine so as to keep the beam as near as possible on the centre of the graticule.

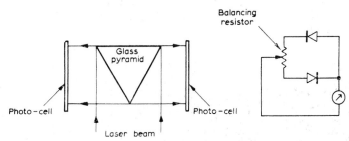

Fig. 9.2. Laser beam alignment head and electronic circuit.

For more precise work an electronic detecting system must be used which can take the form of four symmetrically placed silicon photo-cells[157]. Diametrically opposed cells are linked so as to produce a null reading when each is illuminated by equal intensities. The beam is centred when two null readings from each diametrically opposed pair are obtained simultaneously. The device must be balanced initially by means of ballast resistors incorporated in each circuit (see fig. 9.2) to compensate temperature differences and the differing characteristics of each detector. Each cell must be accurately positioned with respect to the others. This can be achieved by allowing the beam to fall on a reflecting pyramid which can be made of glass to a high standard. The cells are then placed around the pyramid as indicated in fig. 9.2. It is important that the base of the pyramid is kept normal to the incident beam direction.

141

The use of optical components in conjunction with a laser has provided an enhanced method of alignment. The optical components used have consisted either of a coarse circular diffraction grating[158,159,160] as shown in fig. 9.3 or coarse Fresnel zone plates[161] one of which is shown in fig. 9.4. These are fabricated on plane surfaces and the surface positioned so as to be normal to the laser beam.

Fig. 9.3.   Circular diffraction grating.

When a beam of light is incident normally on a circular diffraction grating the effect is to produce, on the far side of the grating, a circular fringe pattern in any plane parallel to the grating. The centre of this fringe pattern defines a straight line which can be used for continuous alignment in conjunction with a graticule consisting of concentric circles.

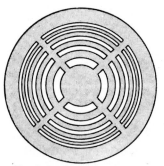

Fig. 9.4.   Fresnel zone plate.

There may be situations where a number of points need to be aligned. A circular Fresnel plate can be used which consists of concentric annuli of varying width as shown in fig. 9.4.

In this case the incident laser beam is diffracted almost entirely to a single point. By using a succession of such zone plates, a series of points can be generated which all lie on the same straight line. This method has been used to construct a 160 m long alignment bench at the National Physical Laboratory. A similar method using orthogonal straight line gratings was used in the alignment of a 3 km long linear accelerator at Stanford University where an accuracy of $\pm \frac{1}{2}$ mm was required over the entire length[161].

In some instances the inclination of some plane is of as much importance as its position. By fixing a plane mirror to the workpiece the laser beam may be returned along its original path and its position at the laser monitored by a photocell array. Clearly the longer the distance between the laser and the object under examination, the greater the sensitivity although the range of angular movement will be reduced.

The accuracy of all these methods relies on the laser producing a beam which has a constant path of propagation. This will depend on the following factors :

1. A constant relative position of the laser mirrors.
2. Stability of the laser structure.
3. Stability of the base on which the laser is mounted.
4. Lack of atmospheric turbulence[162].

The requirements arising from the first three factors can be met to some extent by careful design. Beyond choosing the best time of day for making observations the only answer to atmospheric turbulence is evacuation of the optical system. This was in fact done in the NPL and Stanford alignments mentioned earlier. Such precautions are either inconvenient or impossible in many cases and in these situations atmospheric turbulence is the practical limitation to using lasers (and theodolites) for alignment.

9.2. *Measurement of Distance*

Lasers can be used to measure distances with hitherto unimaginable convenience and accuracy[163].

There are three quite different methods of distance measurement by laser :

(a) interferometric methods
(b) beam modulation
(c) pulse echo.

Interferometric methods are really only applicable at present for distances up to 10 m. Methods (b) and (c) are suitable for measuring distances from 10 m up to millions of kilometres.

We shall now describe these methods and some of their applications.

9.2.1. *Interferometric Methods*

The use of the Michelson interferometer to form interference fringes has been discussed in Chapter 2. It is readily apparent that this instrument can be used to measure distance, or more precisely changes in distance. If either mirror of a Michelson interferometer is moved

143

in a direction parallel to the incident beam, while still remaining normal to it, then the interference pattern will change. A movement of one fringe is caused by the mirror moving through a distance equal to half a wavelength of the light source. By a movement of one fringe it is meant that the illumination of some point in the interference goes through one cycle, i.e. black through bright to black again.

By counting the number of fringes passing a detector when the mirror moves, and knowing the wavelength of the light in air, it is possible to calculate the distance through which the mirror has moved normal to the incident beam with great accuracy over a considerable distance. This method of measurement is not new but until lasers became available was severely limited by the quality of the light sources available. With non-laser sources, coherent lengths are restricted to a few tens of centimetres with a consequent limit on the range of distances measurable.

In practice the conventional Michelson set-up using plane mirrors is not used for two reasons. In the first place light from the laser would be reflected back into the cavity and this would result in an undesirable modulation of the laser output. Secondly plane mirrors are extremely difficult to align.

The commercial systems available use two different arrangements of the optical components as illustrated in figs. 9.5 *a* and 9.5 *b*.

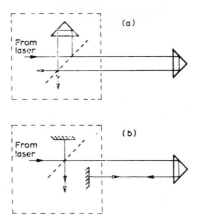

Fig. 9.5. Two practical optical systems for measuring distance by fringe counting.

In each case the fixed part of the instrument is indicated by the components inside the dotted line while the corner-cube prism is fixed to the component whose movement is required. The arrangement shown in fig. 9.5 *a* has the advantage of being easily aligned as two corner-cube retro-reflectors are used. These have the property[164,165,166,167] of reflecting the beam back along a direction parallel to its incident path. The lateral displacement involved is utilized to avoid returned light entering the laser.   Figure 9.5 *b* shows a similar arrangement which has the advantage of being less sensitive to the effects of lateral shear caused by the corner-cube having any sideways motion.   It also has the further advantage of being twice as sensitive because the third mirror produces twice the fringe shift for the same movement.   As a plane mirror has to be used in order that the returned beams superimpose, the alignment is more difficult and returned light entering the laser must be avoided by the use of a polarizing system incorporating a quarter wave plate.

In a conventional Michelson interferometer a change in mirror position of 1 cm causes a movement of nearly ten million fringes. These must obviously be counted electronically by means of a photocell detector[168].

It will be observed that two fringe patterns are formed in the arrangement shown in fig. 9.5 *a* ; these are used to avoid ambiguities caused by vibration.   If only one fringe pattern were detected it would not be possible to determine whether the mirror was moving towards or away from the instrument.   By detecting both patterns in conjunction with a phase sensitive detector the final distance, usually indicated on a digital display, includes the effects of direction reversals caused by vibration or shock.

The speed in which a measurement can be taken depends on the velocity of the mirror.   This is limited by the maximum frequency response of the detector.   In practice movements of 10 cm s^{-1} are possible on commercial instruments.   3 m s^{-1} is available in the laboratory, the latter corresponding to a maximum frequency of 600 MHz.   The amount of shock and vibration which can be tolerated before measurements are disrupted is clearly also dependent on detector frequency response.

In theory the maximum distances measurable with these techniques depend on the coherence length of the laser used.   Providing the laser is stabilized the coherence length considerably increases on changing from multiple axial mode to single axial mode operation and distances of hundreds of kilometres could be measured if it were not for air turbulence which can make the path lengths in each arm of the interferometer vary to such an extent that no spatial coherence exists between the interfering beams and so no fringes are formed.   Air turbulence results in great difficulty in measuring distances greater than 10 m.

With electronic methods of counting fringes it is possible to measure changes of 1/10 of a fringe[169].   Thus a 10 m path length using visible laser light an accuracy of a few parts in 10^8 is possible.   The realization of this accuracy will depend on knowing the wavelength of the laser to the same accuracy.

Under normal conditions a single axial mode output is easily obtained by decreasing the length of the cavity until the axial mode separation is greater than the Doppler width of the gain curve[170].   In this situation only one axial mode can exist under the gain curve at any one instant.

The presence of temperature fluctuations and microphonic disturbances will cause the laser mirrors to vary in separation thus in turn varying the wavelength and intensity of the output. So although single axial mode operation is obtained providing great coherence length it is only possible to state that the wavelength of the output is at some value inside the Doppler linewidth. If for a helium-neon laser the Doppler width is 2000 MHz then for a short unstabilized laser the wavelength is known only to an accuracy of 2×10^9 Hz in 5×10^{14} Hz, i.e. a few parts in 10^5.

In order to realize the full potential of interferometric measuring systems it is clearly necessary to stabilize the laser output. This can be achieved to better than 1 part in 10^7 by piezo-electric vibration of one of the laser mirrors[171] as described in Chapter 5.

There is another form of interferometer suitable for measuring distances which is depicted in fig. 9.6.

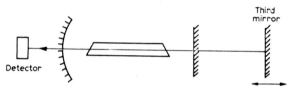

Fig. 9.6.   The third-mirror interferometer.

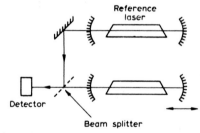

Fig. 9.7.   The beating of two laser outputs to measure mirror movement.

In this system, referred to as the third mirror system and discovered at the Services Electronics Research Laboratory[172], an additional mirror is placed outside the laser cavity. This third mirror reflects light back into the cavity. For certain positions the mirror reflects light back into the cavity so as to be in antiphase with the light inside. The output of the system is consequently of minimum intensity in this situation. By moving the third mirror $\lambda/2$ along the optical axis, the returned light is now in phase and the output is a maximum. By counting the fringes the distance moved by the third mirror can be determined.

146

Measurement of very small movements can be made by interfering two laser beams[173]. Figure 9.7 shows how the output from two lasers is combined at a photomultiplier detector. If one laser is stabilized and a mirror of the other is caused to move, then a change in the beat frequency obtained at the detector will be observed.

We have

$$\frac{n\lambda}{2} = L \tag{9.1}$$

where n is the number of wavelengths in the laser cavity. This number defines the axial mode under consideration. λ is the wavelength and L is the length of the cavity.

$$\therefore \qquad \frac{nc}{2\nu} = L \tag{9.2}$$

∴ differentiating equation 9.2 and ignoring the negative sign :

$$\frac{dL}{d\nu} = \frac{nc}{2\nu^2} \tag{9.3}$$

and hence using equation 9.2

$$d\nu = \frac{\nu}{L}dL \tag{9.4}$$

Thus assuming that frequencies of a few tens of cycles are detectable (an audio frequency) it should be possible, in principle, to detect a mirror movement of 10^{-13} cm. In practice, thermal and microphonic effects as well as changes in the reference laser would make such measurements very difficult. The reference laser can be stabilized to hold the drift to ± 5 MHz per day and thus a 100 MHz beat could be detected with 5% accuracy. For a 10 cm laser tube this would correspond to a mirror movement of 2×10^{-6} cm.

9.2.2. Beam Modulation

A well established method of distance measurement is to modulate a signal which, after transmission and reflection off a target, is detected. The change in phase of the impressed frequency is a dependent of the distance the wave has travelled. By modulating at a number of different frequencies an unambiguous determination of distance is obtained[174,175].

Early devices based on this principle used r.f. modulation of microwaves or microwave modulation of incoherent sources. The use of a modulated laser beam enhances the performance of such a system for the following reasons :

(a) The high frequency of the visible light enables a modulated signal of higher frequency to be used hence increasing resolution.

147

(b) The narrow bandwidth of the laser output enables high discrimination against stray light, thus enabling use in daylight with increased signal to noise ratio.

(c) A high degree of selectivity is possible owing to the small divergence of the laser beam and so different parts of the target can be examined.

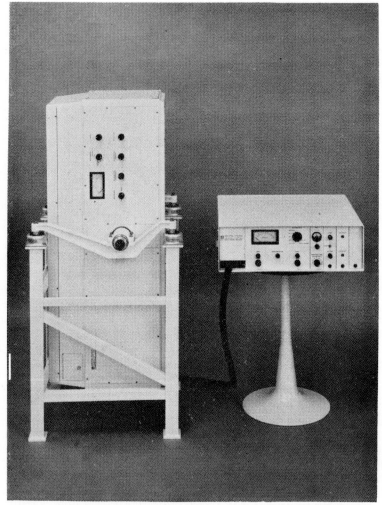

Fig. 9.8.   Laser geodolite.   (Photograph by courtesy of Spectra-Physics Ltd.)

Figure 9.8 is a photograph of a commercially available laser geodolite. In this particular instrument five different modulating frequencies can be applied to a helium-neon laser beam, the highest at 50 MHz enables a resolution of 1 mm to be obtained up to 1 km or 1 in 10^6 at greater distances. With cooperative targets, i.e. those which incorporate a retro-reflector, the maximum range in clear air in daylight is 64 km and at night 80 km. For airborne use the target is usually not cooperative and ranges in clear air are reduced to 3 km in daylight and 5 km at night with accuracies of 2 cm or 1 in 10^4 whichever is greater.

Clearly there are many applications for such instruments. We shall mention two as being of particular interest. The first employs an airborne geodolite working continuously to obtain a terrain profile. Figure 9.9 shows the profile obtained by an instrument installed in an aircraft flying over an urban area.

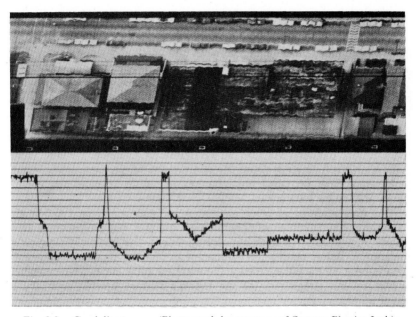

Fig. 9.9. Geodolite trace. (Photograph by courtesy of Spectra-Physics Ltd.)

The black line in the photograph shows the actual path taken by the aircraft. Variations in the height of the aircraft above the ground are compensated for by a barometric pressure recorder.

Another interesting use of the laser geodolite is in the surveillance of dams. Under the weight of water the wall of a dam may well distort by several millimetres. Permanent installation of a geodolite and fixed retroreflectors on the dam wall enable continuous and accurate observation to be made.

149

9.2.3. *Pulse Echo*

Up to this point all the distance measurement devices described have employed continuously operating gas lasers in which use has been made of high coherence or small beam divergence. The use of high power pulsed lasers is also possible in radar type systems where the time interval between transmission and detection is obtained. By multiplying half this time by the velocity of light, whose variations with refractive index, pressure, temperature and humidity are well known, the distance of the target may be ascertained.

Such systems are not novel but have been improved considerably with lasers. Q-switched ruby lasers enable hitherto unimaginable powers to be obtained in very short pulses. The shorter the pulse the more accurately can the distance be determined. The narrow linewidth enables high discrimination against unwanted light to be obtained hence increasing signal-to-noise ratio and range.

A particularly interesting application of pulse echo techniques is the determination of the distance between the earth and the moon. Until about 1957 conventional astronomical techniques such as optical parallax measurements enabled the moon's distance to be established to an accuracy of $\pm$ 3·2 km. More recently by measuring the transit time of a radar pulse a higher accuracy of $\pm$ 1·1 km has been obtained.

The advent of the laser opened up the possibility of an even more accurate determination which would provide information not only on the absolute distance but also on variations in distance. These variations enable information about the distribution of mass inside the moon, continental drift on earth and changes in the location of the earth's north pole to be ascertained.

The first attempt at laser measurement was made in 1962 at the Massachusetts Institute of Technology using a one millisecond laser pulse[176]. Results were disappointing due to the relatively long pulse duration. Better results were obtained in 1965 by Russian workers[177] who used a 50 ns pulse from a Q-switched ruby laser. However the curvature and irregularity of the moon's surface limited accuracy to about 200 m.

The most accurate way of carrying out pulse echo ranging is to use a cooperative target. The amount of light returned can be greater by two orders of magnitude and higher accuracy is obtained by observing the returned beam from a selected point on the target. Cooperative ranging of the moon was possible after the astronauts in the Apollo 11 mission of 1969 placed an array of corner cube reflectors on the lunar surface[178,179]. Figure 9.10 is a photograph of the retroreflector on the surface of the moon.

The array actually consists of 100 individual corner cubes each 4 cm in diameter, the latter dimension being determined by considerations of thermal stability and diffraction. If the corner cube is too small, the returned beam will be very wide by the time it returns to earth and

150

Fig. 9.10. Laser retroreflector on the surface of the moon. (Photograph by courtesy of N.A.S.A.)

insufficient illumination of the detector might occur. On the other hand, if the corner cube reflector is too large the rotation of the earth may cause the returned beam to miss the detector altogether particularly if the laser and detector are located coincidentally. This can happen because the light takes as long as 2·5 seconds to travel from the earth to the moon and back to the earth again.

The 4 cm diameter corner cube reflector causes the reflected beam to be 16 km in diameter by the time it returns to earth.

The transmitting and receiving unit was located at the Lick Observatory at Mount Hamilton in California where the 120 inch telescope was used in reverse to expand the pulse from a ruby laser. The pulses had

151

an energy of 8 J and lasted 10 ns. If the pulses were not expanded they would have been 480 km in diameter by the time they had reached the moon. By passing the pulse through the 120 inch telescope, the diameter on the moon was reduced to about a mile. In fact this is about the practical limit, nothing being gained by using a larger aperture owing to air turbulence. It is obvious that a great deal of light was wasted. In fact each pulse contained 10^{20} photons of which only about 25 were collected after transmission.

Using this technique the earth-moon distance has been determined to an accuracy of $\pm$ 15 cm.

Fig. 9.11. Atmosphere pollution being measured by a pulse echo laser system. (Photograph by courtesy of Laser Associates.)

Pulse echo techniques are also applicable to the measurements of much shorter distances and a number of systems have been built to measure the height of an aircraft above the ground. One of the best of these systems was developed at the Services Electronics Research Laboratory[180] and uses a compact gallium arsenide laser which gives 10 W peak power pulses lasting 30 ns at a pulse repetition rate of 15 Hz. The beam is transmitted via a lens which is coaxial with a large aperture receiving mirror, the returned beam being collected and focussed onto a detector. Heights of up to 300 m can be measured to an accuracy of 1·5 m. Altimeters of this type obviously cannot be used in bad weather or conditions of poor visibility, but they can be extremely useful in checking radio altimeters which are often unreliable over some types of terrain.

152

Much useful information about the atmosphere can be gained by a pulse echo technique which, in this context, is sometimes known as ' lidar ' : *light detection and ranging*[181,182]. By measuring the amount of back-scattered light and its position relative to the laser, the presence of air turbulence[183,184], which may be an aircraft hazard, and atmospheric pollution[185] can be detected. Figure 9.11 shows a pulsed neodymium laser being used to monitor pollution over the industrial region of the Ruhr valley in Germany.

Lidar may also be used to monitor slant visibility in the vicinity of airports and can help pilots to judge whether conditions are good enough for a safe landing.

9.3. *Measurement of Velocity*

The use of the Michelson interferometer to measure distance has been described and it was explained how the speed with which the mirror may be used is limited by the response of the detector. This is because the detector must respond to a rapidly changing illumination caused by the fringes crossing the detector. In fact the speed v at which the mirror moves and the frequency f at the detector can be easily established. If the mirror moves a distance x in t seconds, $2x/\lambda$ fringes will pass the detector, i.e. $2x/\lambda t$ fringes per second, thus

$$f = \frac{2v}{\lambda} \qquad (9.5)$$

where $v = x/t$ is the mirror speed. As an example if the mirror moves with a speed of 1 ms^{-1} then $f \approx 400$ MHz.

Another way of considering this process is to regard the frequency of the light returned by the moving mirror to be shifted by the Doppler effect and then mixed with light of unaltered frequency to produce a beat frequency. Doppler calculations show the light reflected off the moving mirror to be changed in frequency by $2v/\lambda$. Clearly, therefore, it is possible to measure velocity by means of a Michelson interferometer. This method is also referred to as an optical Doppler radar.

This technique for measuring velocity is very useful where the object required is either hot or fragile, so precluding the use of contact methods.

In practice the object is often moving in a direction parallel with its length. In this situation the laser beam has to be incident at some angle, α, to the normal to the direction of the movement in which case the frequency shift becomes

$$\Delta f = \frac{2v \sin \alpha}{\lambda} \qquad (9.6)$$

Under these conditions a reliable measurement of the frequency shift would depend on a constant value of the angle α. As this is usually unlikely, a modified arrangement is used as shown in fig. 9.12.

Here two laser beams are incident simultaneously on the target at different angles. The angle between the beams is constant and equal to $\pi - 2\alpha$. Provided that the bisector of this angle is normal to the direction of motion of the target the same Doppler frequency will be obtained for each beam. This happens because the radar cannot determine whether the object is moving towards or away from the light source. A chopper is used to look alternately at each beam and if there is any difference in frequency, a servo adjusts the beam direction so as to produce a null reading. In this way the angle α can be maintained at a constant value.

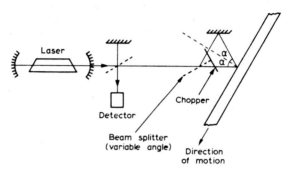

Fig. 9.12. Doppler radar system with compensation for orientation changes in the target.

Such a system has been used successfully to measure the velocity of a hot aluminium extrusion at speeds varying from 1–250 cm s^{-1} and red hot steel bars on billet mills[186]. Another interesting possibility is the measurement and control of the rotation of astronomical telescopes where one rotation of the instrument in 24 hours requires very slow movements to be measured.

9.4. *Measurement of Rotation*
Gas lasers have been used to detect and measure rotation rates relative to an inertial frame of reference. This is in fact angular movement with respect to the fixed stars and therefore suggests an application to inertial navigation. Severe technological problems have limited full exploitation of the method and it cannot yet be considered as a practical application of the laser. However, the interesting properties of devices designed to measure rotation, which are known as ring lasers, have attracted considerable attention and justify discussion.

An optical method of sensing rotation had been demonstrated long before the invention of the laser in the classical experiments of Sagnac[187] in 1911 and Michelson-Gale[188] in 1925. The ring laser rotation sensor is based on these early experiments and is a good example of the influence which the laser can have on optical methods.

154

In the Sagnac experiment light from an arc lamp is filtered and collimated and passed through a beam splitter. The two beams are then reflected round a ring mirror system but in different directions as shown in figure 9.13. On recombination at the beam splitter the two beams interfere and the mirrors are adjusted to show the fringes indicating the phase differences. This is a modification of a simple 2-beam interferometer. If this apparatus is now rotated about an axis perpendicular to the plane of the beams, a fringe shift is observed which is proportional to the rotation rate. If the rotation is clockwise then the light beam travelling in that direction has a longer distance to travel than when the apparatus is stationary and the beam directed in the opposite direction has a shorter distance to travel. Since the velocity of light is the same for both directions there is a difference in transit time for each beam which is indicated by the fringe shift. The sensitivity of this system is severely limited because the fringe shift is proportional to the path lengths traversed and the Sagnac ring with sides of only 30 cm produces a very small fringe shift for a rotation of a few revolutions per second.

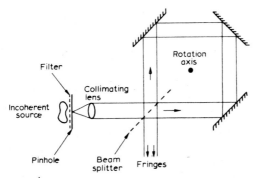

Fig. 9.13. The Sagnac and Michelson-Gale ring interferometer.

The Michelson-Gale experiment achieved greater sensitivity by considerably increasing the size of the ring formed by the mirrors. The beams were passed down evacuated pipes and had to traverse the perimeter of a rectangle 250 m by 125 m in size. Such a large apparatus obviously could not be rotated but a fringe shift due to the rotation of the earth was detected. An independent fringe shift was also observed on a small ring inside the larger ring and the constant proportionallity of the measurements confirmed that the shift was due to the rotation of the earth and agreed with theoretical predictions.

These two experiments, although primarily concerned with an investigation of the effect of rotation on the propagation velocity of light, demonstrated an optical method of sensing rotation rate. However,

155

the large size of the ring necessary to produce measurable fringe shifts made the system impractical. The ring laser rotation sensor improved the sensitivity of these early experiments by several orders of magnitude with a ring area some 200,000 times smaller than that used in the Michelson-Gale experiment.

The ring laser rotation sensor depends upon a frequency difference between two laser oscillations. The two oscillations vary in frequency by virtue of the rotation of the ring and the difference in frequency is proportional to the rate of rotation. The narrow line width laser source is situated within the optical ring in contrast to the relatively incoherent source which is external in the Michelson-Gale system.

The idea of using a laser system for rotation sensing was first proposed by Rosenthal[189] in 1962 and the feasibility of a ring laser was demonstrated by Macek[190] in 1963. The basic features of the first experimental apparatus are shown in fig. 9.14. The square ring laser is

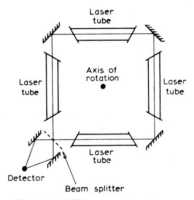

Fig. 9.14.   The square ring laser.

formed by four dielectric coated mirrors at the corners of the square and set at 45° to the optical axis of the helium-neon laser discharge tubes in the four sides. The mirrors are adjusted so that light is reflected from each mirror in turn around a closed loop thus forming an optical ring. The basic requirements for the ring to oscillate are now satisfied ; the discharge tubes provide the source and amplification and the mirrors the in-phase feed back.

The conditions for oscillation are (a) that the gain in the discharge tubes shall exceed the losses at the mirrors and the windows of the discharge tube ; (b) that the mirrors are accurately aligned and the optical path length is an integral number of wavelengths. The wavelength of the oscillations is broadly determined by the transitions in the gas discharge and the wavelength reflectivity of the mirrors. In this case the conditions were selected for the 1·15 μm line of neon. The actual frequency of the oscillation is set by the optical resonator as in a conventional laser and has the same very narrow line width. Traversing identical optical paths and using the same gaseous amplifying medium the clockwise and anti-clockwise oscillations have identical frequencies when the ring laser is stationary with respect to inertial space.

Consider now what happens to the oscillations when this ring laser is rotated in its own plane in a clockwise direction ; the radiation propagating in the same direction as the rotation will have further to go round the resonator to return in phase with itself. The wavelength therefore can be regarded as having to increase slightly in order to fit in the same number of integral wavelengths and the frequency decreases correspondingly. Radiation propagating in an anti-clockwise direction sees an apparently shorter cavity and the frequency of this oscillation increases. The difference between these two frequencies, which is proportional to the rotation rate, can easily be measured and a considerable advantage in discrimination is obtained over the fringe shift measurements in the classical interferometers. Samples of the two oscillations are transmitted through one mirror of the ring and the two beams are superimposed on a photo-detector which has its output displayed on an oscilloscope. This output will then be found to be modulated at a frequency equal to the difference between the two oscillating frequencies. When the rotation rate is such that the difference frequency is in the audio range, the detector output can be amplified to drive a loudspeaker and so produces a striking demonstration of an audio beat note between two optical frequencies of about 4×10^{14} Hz. The lowest difference frequency observed with this first ring laser was 500 Hz which is a discrimination of 1 in 10^{12}. The relationship between the beat frequency and rotation rate can be shown[191] to be of the following form :

$$\Delta f = \frac{4A\omega}{\lambda P} \tag{9.7}$$

where ; Δf = difference frequency between counter rotating oscillations (in cycles)

ω = angular rotation rate (in radians per second)
A = the area of the ring
P = the perimeter of the ring ⎫ in the same units.
λ = the wavelength of the oscillations ⎭

Note that the sensitivity of the ring is determined by the A/P factor and therefore a large ring is desirable although it would require greater stability. The expression is valid for any shape of ring laser : a triangle is a convenient and economical shape and has the advantage that when it is adjusted to oscillate the beams are automatically coplanar. Rotation

157

of the ring about its centre is not essential because off-axis rotation is equal to a rotation plus a translation.

In practice the behaviour of the ring laser is not always so straight-forward as has been described. The ways in which the performance departs from prediction will become apparent if we consider the characteristics of an experimental ring laser.

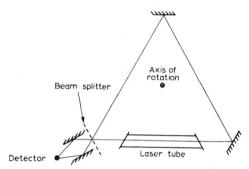

Fig. 9.15. The triangular ring laser.

Fig. 9.16. A ring laser. (Photograph by courtesy of S.E.R.L.)

Figure 9.15 is a diagram of a large triangular ring laser with a peri-meter of 2.6 metres and figure 9.16 a photograph of the device. The oscillating wavelength in this ring is 6328 Å so there are approximately 4·1 million wavelengths around the ring. A rotation of one revolution per hour, which we can consider as an input, produces an output or beat frequency of 1·4 kHz. This is equivalent to about 5×10^6 beat cycles in one revolution or 4 cycles per arc second. If the rotation is reduced to a lower rate, a non-linear effect is observed followed by a disappearance of the beat frequency signal. This occurs, for this design of ring, at a rotation rate of about one degree per minute corresponding to a difference frequency, Δf, of 250 Hz. The reason for this loss of signal is that the two close resonant frequencies of the rotating ring laser pull together and 'lock-in' to the same frequency. A similar phenomenon is observed with two closely coupled electrical oscillators. One reason for cross coupling in the ring laser is that light from one oscillation is scattered at optical surfaces back into the other oscillation[192]. The degree of coupling and hence the locking rate varies from one ring to another. The locking rate is also dependent upon the size of the ring and on the wavelength used. When all possible practical steps are taken to reduce coupling, there remains a possibility of coupling arising from the interaction of the gas molecules in the gain medium.

Some methods of avoiding the locking phenomenon have been investigated[193]. If a constant bias can be applied to one of the oscillations then the resonant frequencies can be kept sufficiently far apart to avoid locking. One way to achieve this is to apply a constant rotation to the ring which is independent of the rotation rate to be measured. This constant difference frequency can be subtracted from the observed beat frequency which is then always above the locking rate. This solution introduces problems of stability, accuracy of imposed bias rotation and so detracts from the relative simplicity of the basic ring laser.

Another method of imposing a constant bias is to introduce into the resonant cavity an optical element with a refractive index which is dependent on the direction in which the radiation is propagating. A Faraday cell, where the refractive index difference can be controlled with a magnetic field, can be used. The optical path length around the ring in one direction is now different from the other thus producing a bias away from the locking rate. This and other bias techniques which require an optical element within the resonator also have stability problems and may be susceptible to drifting.

Work on these tentative solutions to the locking problem have revealed that the many other variable parameters of the ring laser must be carefully controlled. For example, if the perimeter of the ring changes by a fraction of a wavelength, the quality of the beat signal is degraded and the beat count becomes unreliable. Strict temperature control or a servo-mechanism to maintain a constant perimeter is necessary. Another variable is that of gas drift in the discharge tubes ; this may be due to a thermal effect or to drift of the gaseous ions and may impose a random drift bias on the difference frequency. D.C. discharge tubes with currents balanced in opposite directions help to eliminate this effect.

At the present time the performance of the ring laser falls a long way short of that of a good mechanical gyroscope. If a comparable performance can be achieved the advantages of a ring laser as a navigational instrument would be :

(1) No high speed moving parts or long spin-up times as with mechanical gyros.
(2) High sensitivity—theoretically capable of measuring very low rotation rates ; e.g. the rotation of the earth from a stationary ring laser.
(3) Output available in digital form suitable for processing in inertial navigation equipment.
(4) The possibility of lower cost than conventional gyroscopes.

The realization of these advantages awaits the results of further development work.

9.5. *Military Applications of the Laser*

In terms of money spent and human endeavour, the employment of the laser in military equipment must surely be its most important use. The sensitive nature of this subject precluded any mention in the first edition of this book but in the last few years sufficient information has been made generally available for it to be apparent that not only is the subject technically interesting but also that many of the advances made in defence technology would have been impossible without lasers. In this respect the early jibe that the laser was an invention in search of an application has been entirely confounded.

It should be emphasized at the outset that the science fiction concept of using lasers as damage devices whereby the enemy's personnel and equipment are destroyed by burning is unlikely to be realized in the immediate future. The general rule still holds that lasers are expensive and often cumbersome, whereas rifles, guns and conventional ammunition are cheap and portable as well as being more effective. This may not necessarily be true however if the laser were to be used in an anti-ballistic missile role.[370]

Military applications of the laser are included in this chapter because as far as defence is concerned it is in the areas of ranging and weapon guidance that the laser had really come into its own. Work in this field continues at a tremendous pace, particularly in the U.S.A. and because the rapid improvement in techniques and performance make an up-to-the-minute survey difficult, only two systems will be outlined in which lasers have been used to great advantage.

The technology of the modern battle tank has advanced to such an extent since the end of the Second World War that it is now considered highly desirable for tanks to get first-time hits, so destroying enemy armour before the latter can react. The old technique of ranging shots combined with optical rangefinders based on measuring the angle subtended by the rangefinder baseline at the target is therefore inadequate. By installing a Q-switched laser on the turret of the tank its commander can measure his opponent's range with great accuracy and in a time only limited by his own ability to read off figures on a display. Thus even when the tanks of both sides are moving the use of laser rangefinders, stabilized gun platforms and fire-control computers enables high first hit probability to be achieved.

This potential of quick and accurate measurement is being intensively exploited in ranging and target-seeking devices now being fitted to strike aircraft such as the Harriers and Jaguars of the R.A.F.[364,365]. It is now the case that any strike aircraft employing dive-attack-pull up strikes against ground targets is not likely to survive for long before succumbing to the sophisticated ground defence surface-to-air missiles now deployed by many forces throughout the world. Consequently attacks must be made at low altitude and at high speed and so the pilot may only have a few seconds in which to aquire the target, take aim and release his weapons. An essential part of the aiming process is the measurement of range from aircraft to target along the sightline. With a laser this can be carried out almost instantaneously by measuring the time interval between firing the laser and detecting the returned radiation scattered off the target. Furthermore the accuracy of measurement is high as it is limited essentially by the length of the pulse and the discrimination against adjacent objects is good as the divergence of the beam is only a milliradian or so. To get some feel for the numbers involved, take the case of a Q-switched neodymium laser producing a pulse of duration 10 ns to range a target 4 km away with a beam divergence of one milliradian. The time taken for the light to travel from the aircraft's rangefinder and back is only 27 μs, the accuracy of range is the length of the wavetrain which is 3 m and the width of the beam at the target is only 4 m which clearly gives good discrimination. Even at the very high speeds of modern aircraft the range measurement can be repeated 10 or 20 times a second if the laser is well cooled, so providing the pilot or computer with sufficiently updated information.

An excellent example of laser technology applied to military avionics is the LRMTS (Laser Ranger and Marked Target Seeker) being supplied to the R.A.F. by Ferranti Limited. This system, as its name implies, had an additional and very important feature in that the target such as a tank, vehicle or building can be designated by a soldier on the ground with a neodymium laser and binocular sight contained in a portable box and known as a ground marker. Figures 9.17 and 9.18 show front and rear views of the ground marker respectively. On looking through the sight the soldier sees a magnified image of the terrain over which is superimposed a set of crosswires

161

Fig. 9.17. Front view of Ferranti Laser Target Marker/Ranger for the British Army. (Reproduced by courtesy of Ferranti Ltd.)

Fig. 9.18. Rear view of Ferranti Laser Target Marker/Ranger for the British Army. (Reproduced by courtesy of Ferranti Ltd.)

163

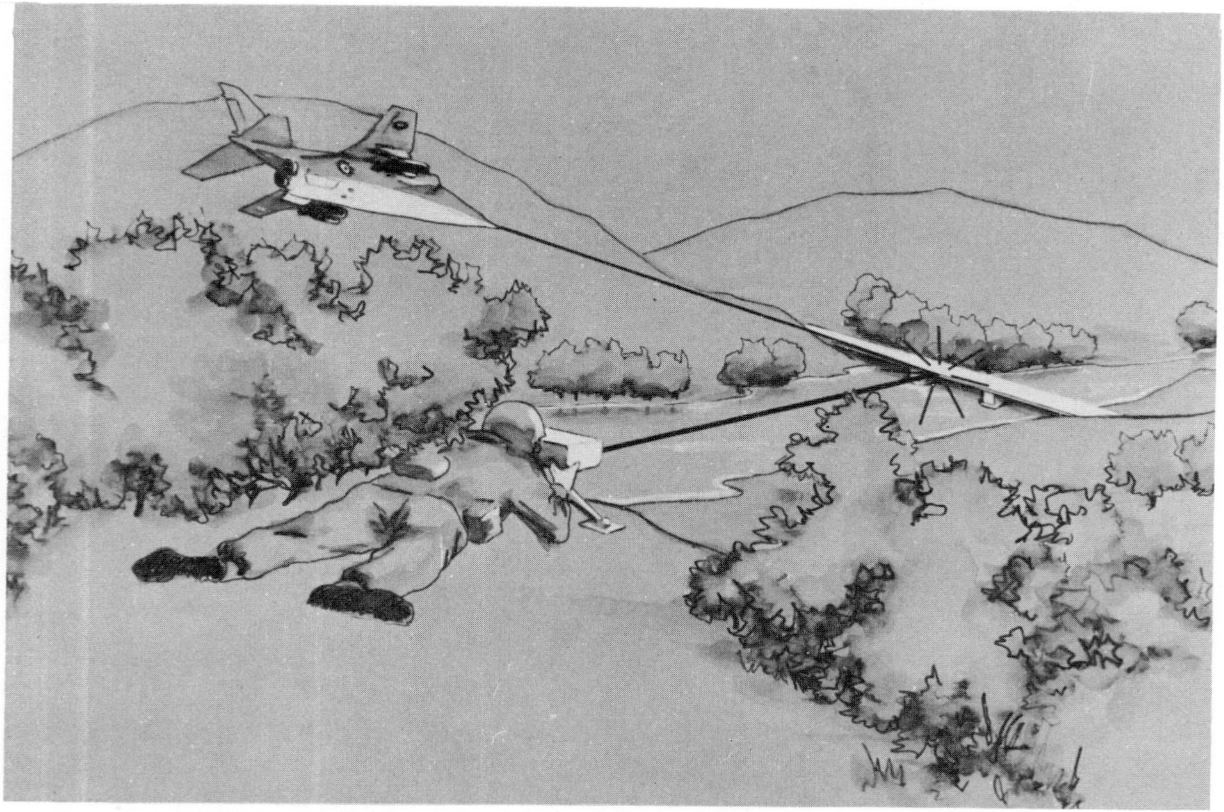

Fig. 9.19. Marked Target Seeking concept. (Reproduced by courtesy of Ferranti Ltd.)

indicating the exact point at which the laser is aimed. The laser spot itself is of course invisible. Figure 9.19 is an artists impression of how the system would work. On the aircraft itself is a gyro-stabilized platform upon which is mounted a reflecting telescope which looks out through a window in the nose of the aircraft. Figure 9.20 shows a Canberra aeroplane with the receiver mounted behind the nose window. The telescope receives scattered laser light from the ground marker via the target because the latter is almost always a diffuse reflector and hence scatters light in all directions to be intercepted irrespective of the position of the aircraft. Mounted at the focus of the telescope is a special position-sensitive detector, the signals from which are used to point the telescope directly at the target. Thus having centred the image of the target the telescope continues to point directly at the target irrespective of how the target or the aircraft moves, always assuming, of course, that the former continues to be designated. Now coaxial with the seeking system in the aircraft is a laser rangefinder employing a water/glycol cooled neodymium laser which is switched on automatically as soon as the target is acquired by the seeker and continually updates range and position information which is fed to the weapon aiming computer aboard the aircraft. The telescope, laser and rangefinder form a compact unit, a photograph of which is shown in fig. 9.21. Finally the pilot's head up display is made to indicate the position of the target and weapon release cues are provided. In this way the pilot can fly a strike mission against a mark on his display—the actual target behind the mark he need not see at all!

Equipment such as that described adds enormously to the efficiency of ground-attack aircraft. It is limited only by weather and visibility conditions and by the pilot's ability to fly close to the ground at night.

Although most strikes by the Americans in North Vietnam were made at medium or high level the effectiveness of laser guided weapons was outstandingly good.[366,367] Modified conventional bombs which were fitted with laser seeking heads. These so-called ' smart bombs ' were thus provided with slight corrections to their ballistic trajectories in order to home onto targets designated by another aircraft separate from the attacking aircraft. Using this simple technique individual spans of bridges or small buildings such as power stations in industrial complexes were selectively destroyed at the first attempt with no damage to the surroundings.

There is little doubt that great advances will be made in the next few years in weapon aiming and delivery using lasers. Already it is known that laser guided howitzer shells have been successfully demonstrated and there seems no reason why laser guidance should not be usefully extended to most forms of ordnance before long.

Fig. 9.20. Ferranti Laser Ranger and Marked Target Seeker installed in a Canberra aircraft of the RAE for early flight trials. (Reproduced by courtesy of Ferranti Ltd.)

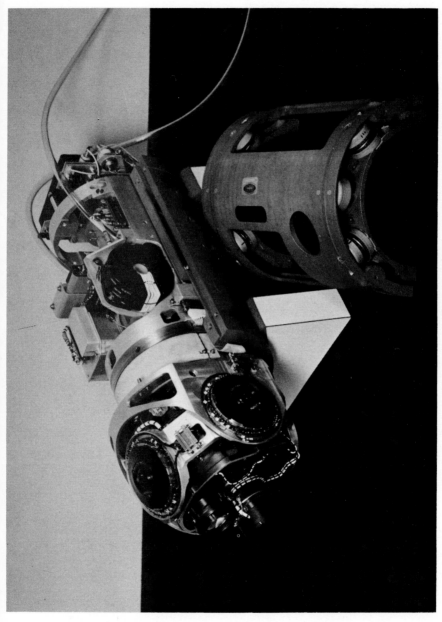

Fig. 9.21. Ferranti Laser Ranger and Marked Target Seeker—internal view. (Reproduced by courtesy of Ferranti Ltd.)

CHAPTER 10

laser communications

MODERN electromagnetic communications systems are based almost exclusively on radio and microwave frequencies. The neglect of optical methods of communication has been due to the adequacy of existing systems, as well as to a number of technical reasons among which the most obvious are difficulties encountered in transmitting a light beam through the earth's atmosphere. Furthermore, most communications systems work on the principle of modulating a carrier frequency with the signal so that one source of radiation can be used to transmit a number of independent signals at different frequencies called channels. In order to maximize the number of available channels it is obviously desirable, therefore, to modulate the carrier frequency at as high a frequency as possible—perhaps to 10% of the actual carrier frequency. The difficulties in modulating a light wave to this degree are considerable.

Carrier type	Frequency range	Usable bandwidth	Approximate number of telephone channels	Approximate number of T.V. channels
LW	30 KHz–300 KHz	10%	3	—
MW	300 KHz–3 MHz	10%	25	—
SW	3 MHz–30 MHz	10%	200	—
VHF	30 MHz–300 MHz	10%	4000	—
UHF	300 MHz–3000 MHz	10%	10,000	10
Microwave	3000 MHz–10^{12} Hz	10%	100,000	100
Optical	5.10^{13} Hz–10^{15} Hz	0·1%	10^8	10^5

Fig. 10.1. A comparison of the number of telephone and T.V. channels which can be transmitted on different types of carrier.

The invention of the laser has provided a potentially ideal source of radiation for an optical communication system. This is not however a sufficient reason for the renewed interest in optical methods which has been brought about for two reasons. First the need has arisen for space communications systems where the necessity for compact, light-weight equipment make the high power density and the narrow beam divergence of the laser very attractive. Secondly, despite atmospheric problems which of course do not arise in space systems, earth optical communications offer an immediate solution to the bandwidth limitations of conventional systems. These limitations are evident from fig. 10.1 which compares different types of conventional communication systems with an optical system, listing their frequency range, usable bandwidth and the number of different telephone and T.V. channels which can be accommodated. It is assumed that a telephone channel requires 5 kHz and a colour T.V. channel with sound requires 8 MHz.

The most complex telecommunication systems today are probably point-to-point systems carrying a variety of channels from telephone to colour television. With these systems the carrier frequency can be as high as 12,000 MHz, which corresponds to a microwave of 2·5 cm wavelength. Higher frequencies are possible but at 24,000 MHz there is an atmospheric absorption band and at the next possible frequency, around 35,000 MHz there are serious difficulties in modulation, transmission and detection.

At present high capacity transmission systems are obtained by duplication of small capacity systems at lower frequencies. Although at the moment this is an acceptable situation it seems probable that with the continually rapid expansion in communications some other solution may have to be found.

The answer may well be provided by optical communications systems incorporating lasers. Optical frequencies range from 10^{13} Hz to 10^{15} Hz so even if only 0·1% modulation can be achieved it is apparent from fig. 10.1 that a vast increase in the number of channels available will be possible.

The basic components of an optical communications system will now be discussed. These are shown in block diagram form in fig. 10.2 and consist of a source, a modulator, some means of transmission and a detector and demodulator.

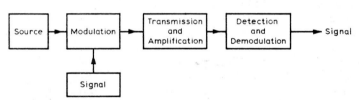

Fig. 10.2. The components of an optical communication system.

10.1. *The Source*

A laser is not necessary for optical communications but it offers a number of overwhelming advantages.

Reference to fig. 4.6 will indicate how the uniphase mode of a laser can propagate over considerable distances with only a relatively small increase in diameter, compared with that which would be suffered by a microwave beam. The angle of divergence α of a Gaussian electromagnetic beam of wavelength λ is given by

$$\alpha = \frac{4\lambda}{\pi D} \qquad\qquad (10.1)$$

Thus for a given initial diameter, D, a laser beam will diverge some 10^5 times less than a microwave beam. Or, in other words, for the same divergence as a 1 cm diameter laser beam, a microwave antenna of diameter 1 km would be required.

The high temporal coherence of the laser and consequent very narrow frequency spread enables narrow band filters to be used so that extraneous light can be eliminated thus increasing the signal-to-noise ratio. The narrow bandwidth and stable frequency of a laser also facilitates homodyne and heterodyne detection which again help to increase the signal-to-noise ratio.

The ideal laser source for communications purposes has the following properties :

(1) high power output.
(2) continuous power output.
(3) robustness and long life.
(4) room temperature operation.
(5) high frequency stability.
(6) good spatial coherence.
(7) high monochromacity, i.e. good temporal coherence.
(8) ease of modulation.
(9) ease of excitation.

The gallium arsenide semiconductor laser has the unique advantage of being easily modulated at up to 10^{10} Hz by simply modulating the d.c. input current[76]. It also has an indefinitely long life. However it suffers from poor spatial and temporal coherence and requires cooling to liquid nitrogen temperatures.

Gas lasers offer by far the most promise as sources for optical communication links and have high coherence, room temperature operation, outstanding frequency stability[129,130,171] (to 1 part in 10^8) and can be modulated internally[42] (although not at very high frequencies). They are however somewhat fragile and of limited life. Nevertheless helium-neon lasers have a life expectancy of over 10,000 hours and argon ion lasers of over 1000 hours. Power outputs are generally not high although 10 watts are available from argon lasers and much higher powers (kilowatts) can be obtained from the carbon dioxide laser. Unfortunately the longer wavelength of the carbon dioxide laser's output results in the beam divergence being inherently some twenty times greater than visible light lasers. Atmospheric absorption at 10·6 μm is considerable and presently available 10·6 μm detectors require cooling.

Solid state lasers can be operated continuously with high power outputs (particularly the neodymium in glass or YAG types) but do not have very good spatial or temporal coherence and cannot be internally modulated. Their lifetime primarily depends on the life of the pumping lamps which are unlikely to last more than 1000 hours.

10.2. *Modulation*

The signal to be transmitted must be impressed on the laser beam by a process known as modulation. Modulation can be achieved in a number of ways, for example amplitude modulation, frequency modulation or pulse code modulation.

10.2.1. *Amplitude Modulation*

The simplest kind of modulation is achieved by modulating the intensity of the laser beam with a mechanical shutter. Such a method is limited in speed and so the bandwidth available is very low. Higher bandwidths can be obtained by modulation of the laser pump. As was previously mentioned the electric current which pumps the semiconductor laser can be modulated at up to 10^{10} Hz with corresponding modulation of the laser output.

With the exception of the semiconductor laser, lasers cannot be modulated at a sufficiently high rate by pump modulation so other methods must be found. Various methods of modulating the output of a laser were described in Chapter 5. The applicability of these methods for communications purposes will now be discussed briefly.

Electro-optic modulators using the Kerr or Pockels effect are often used and have been made to modulate laser beams at up to tens of megahertz[43,194]. However the chief drawback of electro-optic modulators is the high drive powers required. If the capacitance of the modulator is given by C, the voltage on the modulator by V and the bandwidth by f then it may be shown that the pump power P is given by

$$P = V^2 Cf \qquad (10.2)$$

For a longitudinal ADP modulator where $V = 10$ kV and $C = 10$ pF then the power required is 1 kW per megahertz of bandwidth. The length of the modulator known as the interaction length can be increased so reducing the voltage required but a limit is set by the necessity for the light to traverse the modulator in less than half a period of the modulating frequency. Travelling wave modulators have been developed to overcome this difficulty.[195]

The modulator can be placed inside the cavity as discussed in Chapter 5, section 5.3.1. This is advantageous in that the modulation power is reduced by a factor of n where n is the number of passes each photon makes on average up and down the cavity before leaving the laser. As explained in Chapter 4, n is related to the transmissivity of the laser mirrors. Unfortunately this type of internal modulation can be shown[40] to reduce the usable bandwidth by a factor of n. A better system is to use coupling modulation in which case the required modulation power is still reduced by a factor of n while the bandwidth is only reduced to $1-1/n$ of its original width[196].

171

Many other types of modulation can be used although the electro-optic methods have proved most popular, in particular the acoustic modulation methods outlined in Chapter 5, section 5.3.3, has also been widely used[40,197,198,199].

10.2.2. *Frequency Modulation*

In section 5.3.3 of Chapter 5 the Bragg angle acoustic modulator was described. In this system the first order diffracted beam is Doppler shifted in frequency by an amount corresponding to the modulating frequency—in effect single sideband suppressed carrier f.m. modulation. Unfortunately the effect is very weak, the modulated output being 10^3 less powerful than the laser input[40,41].

10.2.3. *Pulse Code Modulation*

An important and extremely promising method of optical communication employs pulse code modulation in which the ability to transmit information is not limited by the bandwidth of the modulator but by the linewidth of the atomic or molecular transition by which laser action occurs.

Briefly, optical communication by pulse code modulation consists of modifying a train of regularly spaced pulses by either transmitting or suppressing any particular pulse. The system is therefore essentially binary in nature. The detector has only to make a decision about the presence of a pulse and not about its amplitude—a great advantage under poor transmission conditions.

Fortunately a train of pulses can be easily obtained from a laser by modulating the gain of the cavity at a frequency corresponding to the frequency separation of the axial modes. This can be done by incorporating a KDP crystal within the cavity. The output of the laser then consists of a train of pulses whose separation is given by $\dfrac{N-1}{N\Delta f}$ where N is the number of axial modes and Δf is their frequency separation. For a large number of axial modes the frequency separation is effectively the inverse of the axial mode frequency separation and the duration of each pulse is the reciprocal of the total frequency width of the laser output. Figure 10.3 indicates these relationships. Using a neodymium doped YAG laser pulses of width as little as 25 picoseconds (25×10^{-12}s) have been generated.

The mechanism by which this process occurs is as follows : when the gain of the laser just exceeds the losses, the axial mode nearest the centre of the gain curve will oscillate. If the laser is being modulated internally at a frequency corresponding to the axial mode separation harmonics will occur on either side of the central axial mode at precisely the frequencies of the adjacent axial modes. As the harmonics and the central axial modes oscillate in phase or, as is often said, they phase lock, the adjacent axial modes will start to oscillate in phase with the central mode as soon as the gain is sufficient. This process is repeated with each of the **axial** modes so that eventually all of them are phase locked. The laser **can** therefore be regarded as emitting a number of different wavelengths corresponding to each axial mode coherently. It is therefore possible for these waves to interfere which they do on propagation from the laser. The net result is a series of pulses which is effectively the propagating interference pattern of all the axial modes.

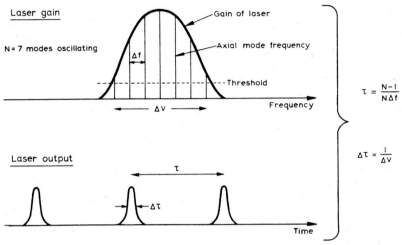

Fig. 10.3. Pulse duration, $\Delta\tau$, and separation, τ, of the output of a frequency modulated laser expressed in terms of the axial mode separation Δf, the number of axial modes oscillating N and the overall bandwidth of the output Δv.

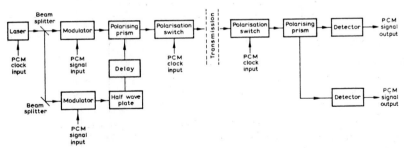

Fig. 10.4. A two-channel time-multiplexed pulse code modulation (PCM) optical communication system.

In a practical system built at Bell Telephone Laboratories[200] a helium-neon laser was modulated at 224 MHz so producing a pulse approximately every 4·5 nanoseconds and each pulse having a width of about 0·6 nanosecond. This output was then coupled into an electro-optic system including a modulator to which is applied the signal in such a form as to modify the pulses. In this case because of the side spacing of the pulses compared with their width it would be possible to time multiplex, i.e. interleave, up to 4 sets of pulses. By using a Nd–YAG laser pulses as narrow as 25–100 ps can be generated and it should be possible to time-multiplex up to 24 channels without interference between neighbouring channels becoming a problem. Figure 10.4 shows the transmission and demultiplexing scheme for a two-channel system.

In this two-channel example the laser output is split into two beams which pass through separate modulators. Each modulator acts as a gate and consists of a lithium tantalate crystal in which the laser beam enters the crystal at 45° to the optic axis and propagates perpendicular to the optic axis. By pulsing the crystal, the beam can be rotated through 90° and so pass an analyser. If no electric impulse is applied the analyser blocks the beam.

The two channels are multiplexed by first rotating the plane of polarization of one channel by 90° with respect to the first and then delaying one beam with respect to the other. Both beams are then passed into a polarizing crystal so that the output consists of two sets of time multiplexed pulses each set being orthogonally polarized. A clock input to a polarization switch rotates the plane of polarization of every alternate pulse before transmission so that a linearly polarized beam is transmitted.

Demultiplexing and detection is achieved by applying the same clock input to another polarization switch. Pulses of different polarization are then separated by a polarizing filter and each detected by an avalanche photodiode directly. The bandwidth of these detectors is quite sufficient to detect 224 MHz signals and more will be said about them later.

Increased capacity could be obtained by using a multiple wavelength laser and pulse coding all the wavelengths. Each wavelength channel could then be isolated by narrow band filters—such an arrangement being known as frequency multiplexing. A further increase could be achieved by using different modes of propagation, i.e. spatially multiplexing.

P.C.M. optical communication systems are still in their infancy and a great deal of development work is required. The theoretical ability of a YAG laser combined with all types of multiplexing to transmit over 10^{14} bits of information per second will no doubt provide an adequate incentive.

10.3. *Transmission*

Probably the greatest obstacle to be overcome in the construction of a viable terrestrial optical communication system is the problem of transmission. The only environment in which unguided transmission is feasible is in space communication[201,202,203]. The earth's atmosphere makes unguided transmission impractical as rain, fog, smoke and haze are almost always present[204,205]. Transmission losses have been estimated at 3–8 dB km⁻¹ for rain, 3–10 dB km⁻¹ for fog and 3–20 dB km⁻¹ for snow[206]. Thermal effects also cause distortion and attenuation. The propagating wavefront can have its direction changed significantly by large thermal gradients and small distortions of the wavefront will occur if the thermal gradients are very local. Atomic and molecular absorption also occur due principally to the presence of water vapour and carbon dioxide in the atmosphere. Figure 10.5 indicates the transmissivity bands or 'windows' of the clear atmosphere over the range of wavelengths at which lasers operate[207].

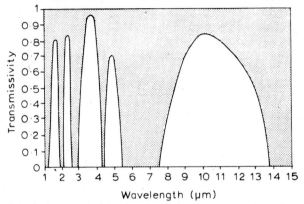

Fig. 10.5. Principal transmissivity bands of the clear atmosphere for radiation of wavelength 1–15 μm.

Clearly most of these effects change the amplitude of the transmitted light beam rather than its frequency and so frequency modulated systems are preferable to those employing amplitude modulation.

Unguided beams have been used to transmit television pictures over distances of up to 20 km in clear weather. However for reliable transmission under all conditions some form of guidance is essential.

Clearly none of the above remarks apply to transmission through space where a major problem will be accurate pointing of the beam.

10.3.1. *Light Pipes*

A variety of different methods for piping laser beams have been suggested and constructed. Essentially the pipes contain dry air isolated from the external environment, although evacuation is preferable as thermal gradients will cause divergence of the propagating beam[208,209].

A practical system has been built[210] consisting of sections of tubing 150 cm long and 2·5 cm in diameter internally coated so as to be highly reflecting. By directing a collimated beam down the tube at an angle of less than 0·5° with respect to the axis, the beam propagates with few

175

reflections and if any do occur, the angle of incidence is so large as to make the losses very small. This system achieved losses of 60 dB km⁻¹ and although it is estimated that losses as low as 1·5 dB km⁻¹ might ultimately be possible, difficulties in construction and coating make the practical achievement of such low attenuations unlikely.

It has been suggested that the problems involved in coating the internal surfaces of such pipes might be avoided by flowing cold gas down a heated pipe. The colder gas in the middle has a higher refractive index and so the gas acts as a continuous lens keeping the beam to the centre of the pipe. Losses could be very low in this system but the difficulty in maintaining an accurate and consistent temperature gradient appears to be a major drawback.

Figure 10.6 shows the principle of a practical system which has a very low loss[211].

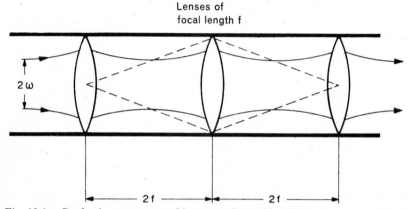

Fig. 10.6. Confocal arrangement of lenses used to guide a laser beam down a pipe.

A series of lenses are used in the confocal mode so as to image each lens on its next but one neighbour, the beam therefore being converged and diverged down the pipe. By ensuring that the width of the beam is very much less than the diameter of the lenses the losses have been kept as low as 1 dB km⁻¹ with the lenses spaced at 80 m intervals. Such a system can become complex and expensive if compensation is made for earth movement and misalignment. Each lens must then be servo-controlled so as to be maintained in an optimum position and so such a system is only really practicable for short distance, high information capacity systems.

Modifications of such a lens system have been built in which the lenses consist, not of glass, but of gas at different temperatures achieved by local heating of the pipe[206,212]. Gas lenses are capable of much less attenuation of the beam than glass lenses. Such systems are very difficult to make however, owing to problems in accurate control of temperature.

176

10.3.2 *Fibre Guides*

Probably the most promising method of transmission is to use fibre optics in which the light pipe consists of a flexible rod of glass a few microns in diameter surrounded by a cladding tens of microns in diameter and of lower refractive index[213,214,215,216]. Most of the energy is transmitted down the central core and because of its small dimension, the pipe can be bent through radii of curvature of as little as 1 cm without adverse effects on transmission. Bundles of such fibres can easily be made to form multi-channel communication cables.

The chief problem with fibre transmission is the high loss associated with materials such as glass and fused silica. Losses in the bulk material may be as much as 100 dB km^{-1} and even higher in fibre form. These losses arise mainly from compounds of iron present in the glass and considerable research effort is being expended in finding ways of reducing these losses. It seems probable that fibres with losses of 10 dB km^{-1} will be made, most of this loss being due to scattering within the medium.

10.3.3. *Amplification*

In any terrestrial optical communication system it will be essential to provide repeater points at which amplification of the light beam can take place. These amplifiers may well take the form of lasers without the feedback normally supplied by the mirrors, i.e. material with an inverted population. Although amplification has been demonstrated at a number of different wavelengths, it is always found that high gain can only be achieved at low power levels. Gas amplifiers[217,218] work continuously with little noise but only at low power with the exception of the carbon dioxide amplifier[219,220,221] which amplifies at high power but with low gain. Semiconductor[222] and solid-state[223] amplifiers have only been operated in a pulsed mode.

10.4. *Detection and Demodulation*

The first component of any detection system will almost certainly be a narrow-band filter to exclude extraneous light. Two types of detection can be used, these are referred to as (1) direct or incoherent and (2) heterodyne or coherent.

10.4.1. *Direct Detection*

If the modulated laser beam is allowed to fall on the detector so that the latter responds directly to intensity variations then direct detection takes place[224]. The modulated beam should be preamplified until the noise in the signal is equal to that in the detector.

The conventional photomultiplier can be used for direct detection providing the modulation is less than 10^8 Hz. This limit is set by the variation in transit time of the electrons between the dynodes caused by a variation in the velocity with which the electrons are ejected from the cathodes. Special types of photomultiplier have been constructed in which the spread in transit times is reduced, these are known as crossed field photomultipliers and can handle signals of up to 5×10^9 Hz.[225]

Another method of demodulation is to use a travelling-wave phototube[225] as illustrated in fig. 10.7.

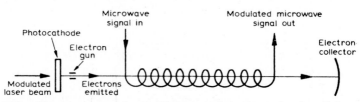

Fig. 10.7. The components of a travelling-wave phototube used to demodulate a laser beam.

The modulated laser beam is incident on the photocathode and the emitted electrons are directed by means of an electron gun towards the collector. On the way they pass through a helical slow-wave structure along which a microwave beam is transmitted. The modulation on the laser beam is transferred via the electron beam onto the microwave which can then be demodulated by conventional methods.

Both the crossed-field photomultiplier and the travelling wave phototube are photoemissive devices and are restricted in use by the wavelength of the incident radiation. They will in fact work from the ultraviolet to the near infrared but for wavelengths longer than 1 μm or so photoconductive detectors must be used.

Photoconductive detectors can be divided into those such as indium antimonide and lead selenide, which require cooling to liquid nitrogen temperatures, and p–n junctions which work at higher frequencies than the 1 MHz to which the ordinary photoconductive detectors are limited. These diodes are reversed biased p–n junctions in which the reverse current is modulated by the incident radiation. By suitable doping, some degree of internal gain may be achieved through ionization at the junction. These avalanche diodes, as they are called, are usually made of silicon or germanium. Unfortunately silicon avalanche photo-diodes do not work at wavelengths longer than 1 μm while the use of germanium only enables radiation out to 1·6 μm to be detected.

10.4.2. Heterodyne detection

Heterodyne detection takes place when the modulated laser beam is mixed with another unmodulated beam and the resulting beat frequency containing the signal detected directly[226]. If the carrier and the mixed beam are of exactly the same frequency homodyne detection is said to take place. Both homodyne and heterodyne methods have the advantage that the theoretical signal-to-noise ratio is twice as great as that for direct detection[227].

178

Due to the previously described difficulties in direct detection in the infrared, heterodyne detection would seem to be suitable for carbon dioxide laser communication systems[203,228,229]. However the local oscillator and the received beam must be accurately aligned and spatially coherent. This requirement demands high standards in acquisition of the signal beam, in transmission and in the optical quality of the components used in the receiving system[230,231].

CHAPTER 11

the laser as a heat source

THE idea of using light as a heat source is not new. In the third century B.C. Archimedes attempted to destroy the Roman fleet by constructing large parabolic mirrors which focussed the rays from the sun to a small spot. Only since the invention of the laser however has a focussed beam of light become a practical possibility as a source of heat in industrial processes. These possibilities are attributable to the spatial coherence of the output, or, in other words, to the ability of the laser to produce a beam of light of very small divergence. This enables the beam to be focussed, usually by means of a lens, into a spot whose diameter can approach the wavelength of the radiation emitted by the laser. Thus the energy contained in a laser beam of a few millimetres in diameter can be concentrated into an area of a few square microns so giving an increase in energy density of at least six orders of magnitude.

If a beam of light diverges by an amount θ and is focussed by a lens of focal length f, then, assuming that the aberrations of the lens are negligible, the diameter S of the focussed spot is given by

$$S = f\theta \tag{11.1}$$

Equation 11.1 only holds when θ is non-zero.

Thus if the power in the beam is W watts then the power density E at the focussed spot is given by

$$E = \frac{4W}{\pi f^2 \theta^2} \tag{11.2}$$

For a beam of uniform intensity distribution and radius R

$$\theta = \frac{1 \cdot 22\lambda}{R} \tag{11.3}$$

and so

$$E = \frac{0 \cdot 86 \ WR^2}{f^2 \lambda^2} \tag{11.4}$$

If the laser is operating in the uniphase mode and the lens radius R is sufficient to exclude only light of intensity less than $1/e^2$ that at the centre then

$$\theta = \frac{2\lambda}{\pi R} \tag{11.5}$$

and so in this case

$$E = \frac{\pi WR^2}{f^2\lambda^2} \qquad (11.6)$$

Comparison of equations 11.4 and 11.6 show that the Gaussian property of the uniphase output of a laser enable higher power densities to be obtained than from uniform beams. However most pulsed lasers and those c.w. lasers giving maximum power have multi-transverse mode outputs and beam divergences are usually between 2 and 20 milliradians. These are an order of magnitude greater than the beam divergence from a uniphase output and equation 11.2 must be used to estimate power densities.

The power densities given in the above equations are derived on the assumption that the light is concentrated uniformly over the focussed spot. This is never the case however in practice. For uniform illumination the intensity distribution at the focus is actually given by a function of the form $(\sin x/x)^2$ and for Gaussian illumination the intensity distribution will actually be another Gaussian. In addition lens aberrations will modify these distributions. In all cases the intensity at the centre of the focussed spot will be greater than at the 'edge'. Nevertheless the equations provide a good indication of the power densities available.

As examples of power densities possible, consider a pulsed ruby laser giving a power output of 10 MW with a beam divergence of 10 mrad. and focussed by a lens of 10 cm focal length. By substitution into equation 11.2, $E = 10^9$ Wcm^{-2}. For c.w. operation the carbon dioxide laser is often used. In this case beams of 1–2 cm diameter are obtained with divergences of a few milliradians. These can be focussed to spots 50 μm in diameter with power densities of several megawatts per square centimetre.

It is apparent therefore that the power densities generated by a laser can be utilized in applications such as the welding, cutting and ablation of materials. These processes will be considered in detail later.

One of the advantages of the laser as a heat source is that the small spot size enables operations to be carried out close to regions or components which are heat sensitive. Rapid cooling of the heated area can be effected as surrounding regions act as a heat sink. It should be remembered that despite the high power densities generated the actual quantities of energy delivered may be quite small and any excess above that required for melting or vaporization can often be easily dissipated.

Laser burning and welding techniques are essentially non-contact methods so these operations may be carried out on encapsulated materials with no possibility of contamination.

The actual processes involved when a focussed laser beam is used as a heat source are not fully understood and most experimental work has been conducted on an empirical basis. No attempt will be made

181

here at a theoretical discussion but significant experimental observations will be described.

11.1 *Welding*

The welding process involves the melting of the edges of separate pieces of material so that they fuse together to form a continuous solid structure on cooling. A particular advantage of laser welding of some metals is that impurities such as oxides are brought to the surface during the welding process with the result that the weld can be stronger than that which would have been obtained by conventional methods.

Argon arc welding equipment can produce power densities of more than 10^4 Wcm^{-2} and oxyacetylene burners up to 10^3 Wcm^{-2}, both many orders of magnitude less than that available from a laser. Electron beam welding gear can, it is true, achieve comparable performance but the necessity for a vacuum environment makes such a system impractical in many cases.

An important consideration in laser welding is that sufficient power must be supplied to melt the material but not to evaporate it[240]. Consequently welding is more difficult with materials such as chromium and tantalum whose melting and boiling points are close together. With these metals careful control of beam power is necessary to make good welds. Materials such as gold, copper and nickel have widely separated melting and boiling points and are hence easier to weld[241].

Difficulty in welding metals can arise from their high reflectivity at long wavelengths and their reduction in reflectivity on melting[232]. Consequently, if vaporization is to be avoided[245], beam power and exposure times are critical particularly when using a carbon dioxide laser[233,234,235,236,237]. One solution to this problem is to supply high power initial breakdown pulses followed by further exposure at reduced power.

Carbon dioxide laser welding of metals is restricted to thicknesses of less than about $\frac{1}{2}$ mm as sideways diffusion of heat in greater thicknesses causes the welds to be unacceptably wide. It has been found though, that if the metal is oxidized before welding, so that the surface is blackened, greater thicknesses can be welded.

Pulsed ruby and neodymium lasers have been used successfully for welding. The pulses must be sufficiently long to ensure that powers are not high enough to vaporize the material. Pulse durations of as long as 10 ms are obtainable from commercial solid state lasers and these have proven to be sufficiently long. Welding speeds with pulsed solid state lasers are limited by the pulse repetition rate, which does not usually exceed one pulse per second. Q-switched carbon dioxide lasers can however achieve repetition rates of several hundred pulses per second.

Figure 11.1 shows some welding data using a carbon dioxide laser[266,267].

Material	Laser power	Material thickness	Weld type	Cutting rate
Stainless Steel	{ 145 W 600 W	250 μm 12 μm	edge edge	1·25 cm s^{-1} 80 cm s^{-1}
Carbon Steel	250 W	300 μm	seam	0·6 cm s^{-1}
Quartz	{ 50 W 50 W	0·5 cm 1 cm	seam seam	0·08 cm s^{-1} 0·016 cm s^{-1}
Polythene	600 W	12 μm	edge	80 cm s^{-1}
Nimonic 90	600 W	25 μm	butt	3 cm s^{-1}
Tantalum	600 W	12 μm	butt	2 cm s^{-1}
Titanium	600 W	50 μm	butt	2 cm s^{-1}

Fig. 11.1.   Welding data for various materials using a carbon dioxide laser.

A considerable amount of information is available on welding wires using a pulsed ruby laser[238,239]. This technique has been used in the micro-circuit industry for welding interconnecting leads to components. In this application, lack of damage to adjacent areas, remote operation and absence of contamination are obvious advantages. Typically, 25 μm thick gold or copper wires can be satisfactorily welded by a 3 ms pulse of average power 10 W.

11.2. *Ablation*

If the power in a focussed laser beam is sufficiently high, the material may be raised above its boiling point and removed completely by evaporation[253]. This ablation process has a number of applications including, machining and vaporization of material for spectroscopic analysis.

When evaporation of material is required pulsed lasers are generally used to achieve sufficiently high power densities[240]. This enables the melting point of the material to be reached rapidly, the definition of the affected area is increased and the depth of affected material decreased which is desirable when heating of the substrate needs to be avoided. On the other hand pulse lengths should not be so short that thermal shock occurs ; this may be a problem with low thermal conductivity materials although the latter characteristic does enable the melting point to be more easily reached.

Hitherto some difficulty has been encountered in drilling holes in diamond because diamond-tipped hardened steel bits and expensive drilling pastes must be used and the process takes a considerable time. By using a lens to focus the beam from a pulsed ruby or neodymium laser holes can be drilled in a few minutes[240,242,243].

Figure 11.2 shows a commercially available laser drill for making holes from 0·1 mm to 0·95 mm in diamond dies. The dies are then used in the manufacture of fine wires by pulling metal through the holes. The system consists of a neodymium laser producing up to 10 pulses a second, each of energy 1 J and a lens which focuses its beam to a fine spot on the diamond die. The latter is held in a metal collar which can be rotated at up to 10 rpm and with an off-centre

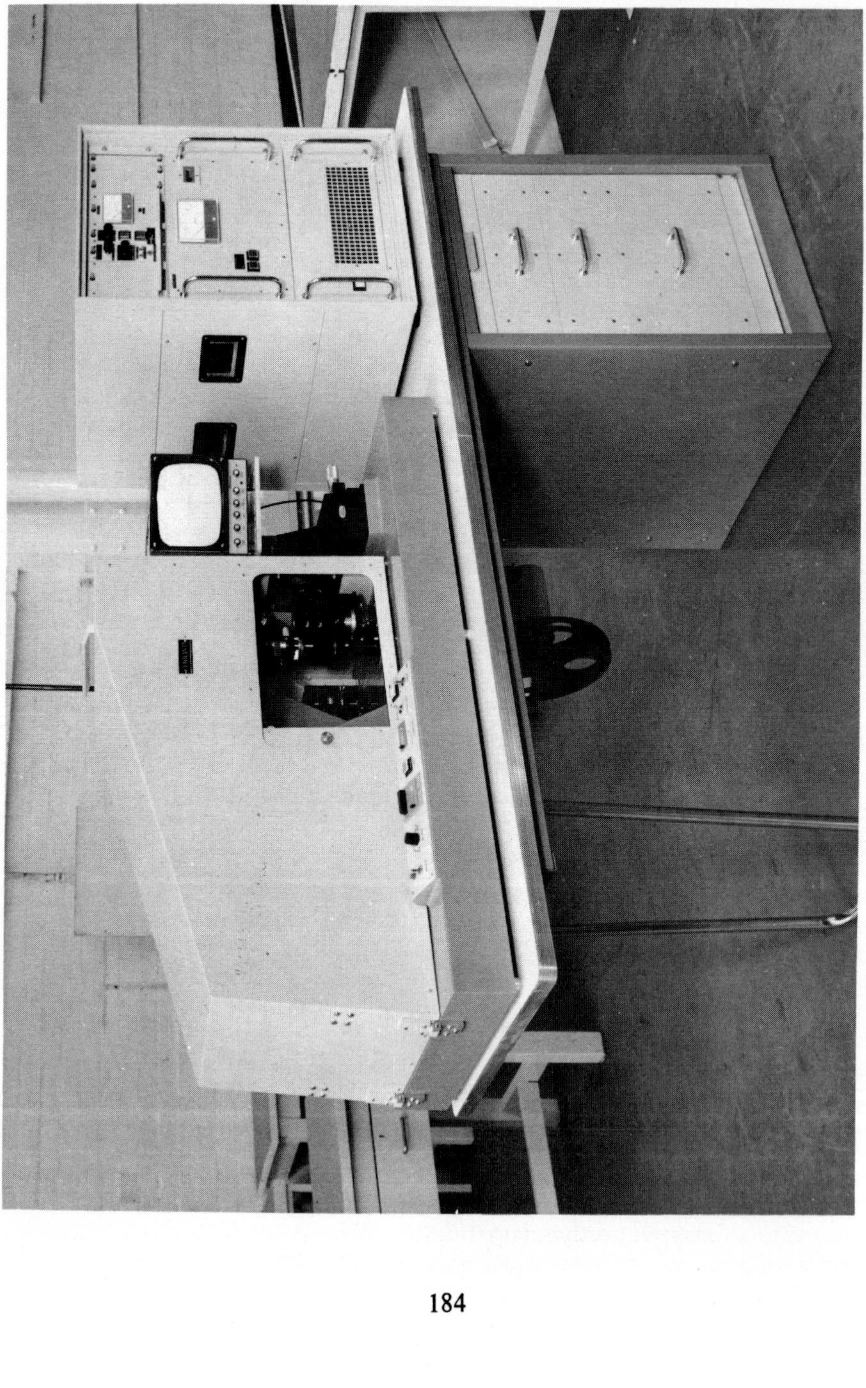

Fig. 11.2. Diamond die driller. (By courtesy of Ferranti Ltd.)

axis of rotation for the larger holes and the operation can be viewed remotely and safely on the TV monitor. By using many pulses it is found that a smooth hole profile can be obtained and the life of the instrument is enhanced. To drill a 0·1 mm diameter hole in 1 mm thick diamond 600 pulses are used but the drilling time is only 2 minutes. The beam focusing lens is protected from sparks and vapour of removed material by automatically passing a transparent melinex tape from a spool in front of the lens as drilling proceeds. An accurate positioning table with x and y movements enables the hole to be positioned to an accuracy of 25 μm in both directions.

Although a focussed beam of laser light is used for drilling, the holes so made can be of almost uniform width and considerable depth. The ratio of depth to hole diameter (the aspect ratio) can exceed 20 to 1 in some materials. These high aspect ratios are thought to be possible on account of reflections back and forth inside the hole as indicated in fig. 11.3

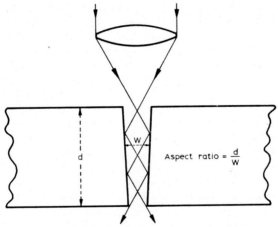

Aspect ratio = $\dfrac{d}{W}$

Fig. 11.3. Hole drilling with a focussed laser beam. Reflection inside the hole enables high aspect ratios to be achieved.

Materials which are brittle like silicon or ceramics and materials which are hard such as tantalum, tungsten, molybdenum, niobium and tool-steel are especially amenable to laser drilling.

Further examples of laser machining include the trimming of resistors where low tolerances are imperative[244,245,246,247]. Encapsulation of thick film resistors often changes their resistivity so by vaporizing some of the resistive material after encapsulation the resistor can be trimmed to be within 0·01% of its nominal value. Tantalum nitride resistors have been trimmed using pulsed neodymium or argon lasers[248]. An important advantage of this method is that trimming can be carried out anywhere in the production sequence.

Capacitors are sometimes made by forming gaps in fine gold films.

Lasers can be used to form these gaps, the fine control possible enables precise components to be manufactured[249].

Lasers have also found a use for scribing fine lines on materials. The silicon wafers on which microcircuits are made must be divided up into individual chips after the array of circuits has been made, each chip then containing an individual microcircuit. The division is usually accomplished by scribing fine lines on the wafer with a diamond point and breaking the wafer along the scribed lines. The diamond scribe frequently leaves rough edges and contaminates the microcircuits thus reducing the yield. It has been found that a considerable improvement can be made to this process by scribing the lines with the focussed beam from a Q-switched neodymium-YAG laser[250,251].

Another application in microcircuit manufacture is in mask fabrication[252]. A helium-neon laser working at $1 \cdot 15 \mu$m has been used to form masks for making interconnecting circuits. The output from the laser, which produces pulses 500 ns long with a peak power of 200 W, is focussed to a spot some 10 μm in diameter. The power density is about 10^8 Wcm^{-2} which is sufficient to quickly evaporate thin films of chromium, tungsten, tantalum and nichrome deposited on borosilicate glass substrates. By keeping the focussed spot fixed and placing the coated substrate on a movable coordinate table a complicated pattern can be cut which can be controlled by means of a tape input to a computer. The advantages of this method of mask manufacture over conventional processes are the convenience of computer control, the elimination of detailed artwork and the ability to produce a set of masks in only a few days.

The drilling, cutting and welding of plastics are particularly easily carried out with a carbon dioxide laser. The $10 \cdot 6 \mu$m radiation is highly absorbed by some plastic materials which contain absorbing compounds such as colouring dyes and their low thermal conductivity enables high local temperatures to be reached quickly.

Another interesting use of the laser in the field of machining is the balancing of gyroscopes[254,255]. Gyroscopes run at very high speeds, up to 25,000 r.p.m. and consequently require very accurate balancing. Conventional balancing methods involve running the gyro up to its operating speed, detecting the point of imbalance, running the gyro down, clamping to prevent damage to bearings and removal of the offending material. This process can take several hours. By trimming away the material with a pulsed laser while the gyro is running, balancing times can be reduced to 10–30 minutes. Furthermore, bearing damage and contamination is avoided as the removed material is swept away by centrifugal forces.

Lasers can help in the field of spectral analysis by absorption spectroscopy[256,257,258]. By evaporating a very small quantity of material with a focussed pulse from a Q-switched ruby laser and providing further excitation by an arc across two carbon electrodes just above the surface of

186

the material, it has been found that only 10 grams of a material are required and that an increase in sensitivity is obtained. As very small amounts of material are sufficient, non-destructive analysis may be carried out over selected regions of a sample and the speed of operation enables rapidly changing objects to be examined.

are many industries in which large quantities of materials of different

11.3. *Cutting*

The cutting of wires and sheets of material is one of the best applications of the laser. The high speed at which cuts can be made and the ease of control of laser beams facilitates numerical control. Dust and waste are reduced and the cutting operation can be almost silent. There are many industries in which large quantities of materials of different shapes are needed, for instance in the manufacture of cutting tools for the fabrication of cardboard boxes[259] and in cutting cloth for making suits.

Carbon dioxide lasers operating continuously can cut almost anything[235,236,237,260] if the beam is focussed and a jet of gas directed at the cutting area[261]. Figure 11.4 shows a diagram of a gas cutting head developed by the British Welding Institute[266]. A beam from a carbon dioxide laser is focussed to a small spot some 500 μm in diameter. At the same time gas at a pressure of 10–15 lbs/sq. in. is introduced concentrically at the focus of the beam.

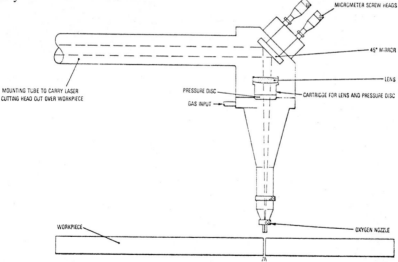

Fig. 11.4. A gas-assisted laser cutting head developed by the British Welding Institute.

If the introduced gas is oxygen, metals can be cut far more easily while at the same time the profile of the cut is improved with less rounding off of edges. It is thought that non-ferrous metals form oxides which have a much lower reflectivity than the pure metal and hence adsorb the

187

Fig. 11.5.   Mild steel plate being cut by an oxygen-assisted carbon dioxide laser.

energy of the beam more easily.   In the case of titanium and ferrous metals exothermic chemical reactions occur which further raise the temperature.

Figure 11.5 is a photograph of a mild steel plate being cut with an oxygen assisted carbon dioxide laser.

Gas assistance is not necessary for cutting materials such as wood and paper but the use of an inert gas or carbon dioxide reduces charring probably by preventing the ignition of gaseous by-products in the vicinity of the cut.   Inert gas jets may also be advantageous in producing clean cuts in other materials by rapidly removing by-products.

Figure 11.6 lists some examples of materials which have been successfully cut using a c.w. carbon dioxide laser together with details of thickness, cutting rates and gas assistance[236,266].

The carbon dioxide laser shown in fig. 7.13 forms the basis of a commercially available cutting unit shown at work in fig. 11.8.   Most materials can be machined and cut with the 450 W available from the laser as the TEM_{00} mode can be focused down to 0·1 mm diameter with a corresponding energy density of 5 MW cm^{-2}.   The exceptions

188

are pure metals such as copper and aluminium which have a high reflectivity at $10.6\,\mu$m. As an indication of the range of materials and advantages obtainable with this particular system, see fig. 11.9, the data for which was kindly supplied by Ferranti Ltd. Fig. 11.10 lists thicknesses of various materials and cutting speeds possible.

Material	Thickness (mm)	Laser power (W)	Power density (MW cm^{-2})	Gas	Cutting rate (cm s^{-1})
Mild Steel	0·5	200	—	oxygen	1
	3	350	0·5	oxygen	1·5
Stainless Steel	0·5	200	—	oxygen	4·4
Titanium	0·6	200	—	air	0·3
	1·5	300	0·25	oxygen	4
Zirconium	0·25	200	—	air	1·5
Nimonic 90	1·5	250	0·4	oxygen	1
Sintered Carborundum	1·6	200	—	air	1·3
Asbestos Cement	6·3	200	—	air	0·04
Silica	1	600	2	none	5
Glass	4	200	—	air	0·16
Perspex	25	200	—	air	0·16
Nylon	0·8	200	—	air	8
PTFE	0·8	200	—	air	10
Leather	3·2	200	—	air	10
Deal	50	200	—	air	1·5
Oak	18	200	—	air	3
Teak	25	200	—	air	0·12
Synthetic Rubber	2·5	600	2	nitrogen + steam	4

Fig. 11.6. Cutting data for various materials using a carbon dioxide laser.

11.4. *Eye Surgery*
A number of medical applications have been suggested for the laser. All utilize the high power densities which the laser can provide and include the treatment of dental decay, the destruction of malignant tumours, the treatment of skin diseases and genetic research by the irradiation of single cells.

Most of these applications, although interesting, and meriting further research, have so far been somewhat inconclusive in their results. A notable exception however has been surgery of the eye for the treatment of detached retinas. The retina, which is the sensitive part of the eye containing the rods, cones and nerve endings which transmit visual sensations to the brain, is normally attached to an outer structure called the choroid. Under some circumstances the retina can become detached from the choroid causing local blindness over the separated areas. The remedy for this defect is to cause a coagulation of the retina and choroid by local heating. In the past this operation was performed using high power xenon arc sources, the light from which after rough collimation was directed at the eye and brought to a focussed spot on the retina. This method has a number of disadvantages such as the need for bulky equipment, exposure times of several seconds during which the eye must remain still and difficulty in focussing the beam to a small enough spot. By using a ruby laser with a portable power supply, smaller areas of coagulation may be made in a few milliseconds a sufficiently short time to make eye movement of no significance.

Figure 11.7 shows a commercially available laser opthalmoscope developed at the Department of Ophthalmology of the Royal Victoria Infirmary, Newcastle and International Research and Development Limited[262].

Essentially such instruments consist of a conventional ophthalmoscope into which is built a ruby laser. A set of crosswires is projected onto the retina so enabling the surgeon to aim the instrument accurately. The power required to form a satisfactory coagulation varies considerably according to the particular area of the eye concerned, therefore the power supply incorporates an adjustment so that the power in each pulse can be gradually increased until the energy is sufficient to attach the retina to the choroid.

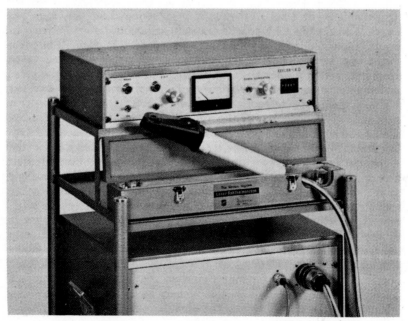

Fig 11.7. Vernon Ingram laser ophthalmoscope control system with laser ophthalmoscope head. (Photograph by courtesy of Keeler Optical Products Ltd.).

190

11.5. *Laser Safety*

The ability of the laser to·be used in the treatment of detached retinas is a graphic indication of the ease with which damage to the eye might be caused by misuse of a laser. It has been reported[263] that for radiation in the visible region of the spectrum the maximum permissible power density on tissues is about 100 mW cm^{-2}. If it is assumed that most visible laser beams will approximately fill the pupil of the eye, then the power density at the retina must be magnified by a factor of 10^5 to account for the focussing action of the eye. This implies that any c.w. laser producing more than $0·1$ μW output is likely to damage the retina if directed at the pupil. Even the smallest c.w. lasers produce two orders of magnitude more power than this.

When using pulsed lasers the energy density is a more relevant parameter and energy densities of 10^{-8} J cm^{-2} at the retina are regarded as the limit for normal pulsed operation. This figure should be reduced to 10^{-9} J cm^{-2} for Q-switched working. Commercially available ruby lasers can provide as much as 10^{-1} J cm^{-2} in each pulse and most solid state lasers give densities many orders of magnitude more than the danger level at the pupil of the eye.

There are many factors such as the wavelength of the laser output, the diameter of the pupil of the eye and the accommodation of the eye which can be taken into account when discussing safety precautions with visible light lasers. However in practice these are somewhat academic as they in no way alter the conclusions to be drawn from the simple energy density figures just presented. It cannot be over-emphasized therefore, that the eye must never be subjected to the direct beam from a laser *or to reflections of the beam*. Particular care is needed when moving or adjusting optical components in a laser system.

The human eye transmits over the range of wavelengths from approximately $0·4–1·2$ μm so the radiation from·a carbon dioxide laser will not reach the retina[264,265], or be focussed, as it is absorbed in the eye. Consequently carbon dioxide lasers are inherently much safer than lasers emitting visible radiation. Nevertheless surface heating effects can occur at the cornea (the front surface of the eye) with dire results on vision. For many years the maximum safe recommended level for men working near blast furnaces has been 10 mW cm^{-2}. As much of this radiation is in the infrared, a safety level of 1 mW cm^{-2} has been suggested for the carbon dioxide laser. With very high power densities severe damage and even perforation of the cornea can occur. It is advisable to wear safety goggles at all times when working with carbon dioxide lasers. This is particularly important in view of the invisibility of the beam especially where reflections are concerned.

Finally, at the risk of stating the obvious, unless special precautions are taken only authorized and knowledgeable personnel should be allowed in the vicinity of a working laser.

191

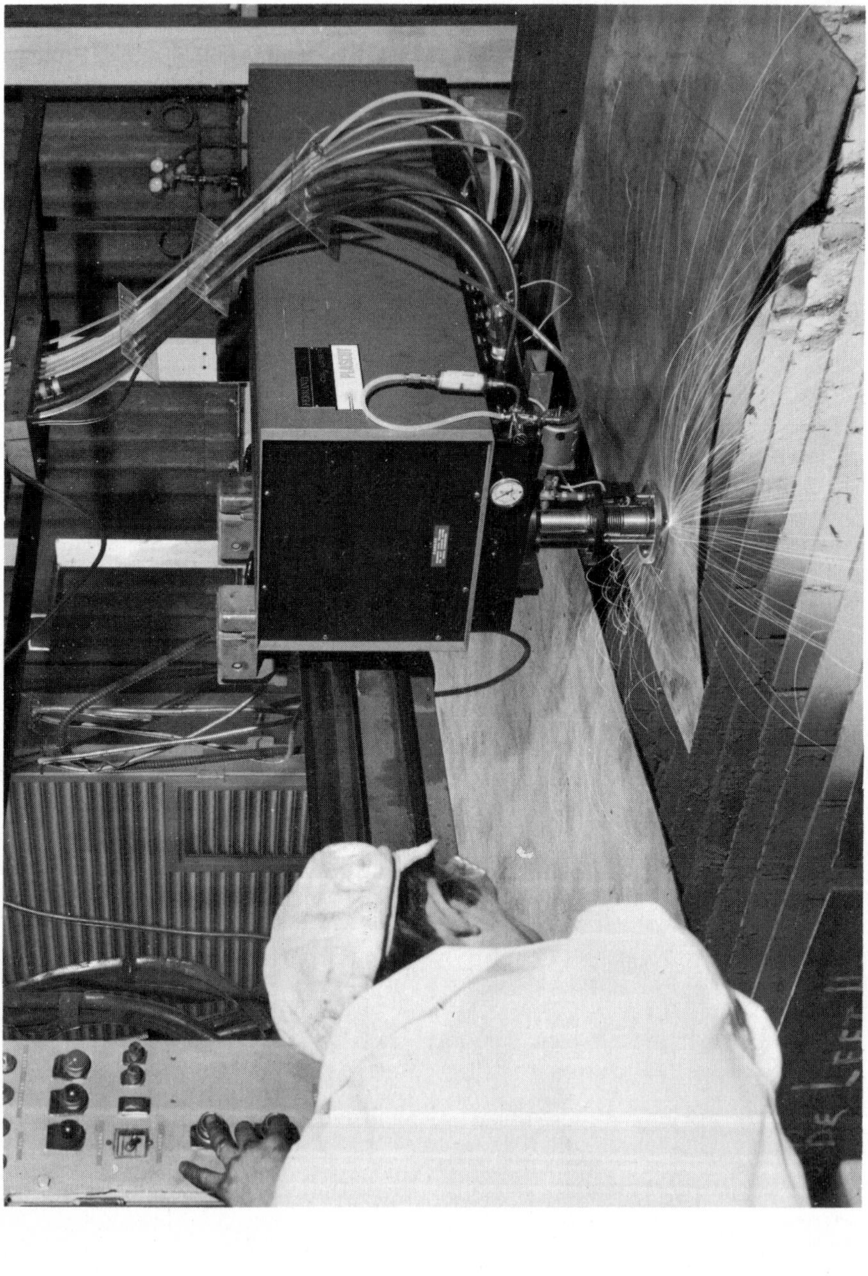

Fig. 11.8. Ferranti MF.400 Industrial laser system. (By courtesy of Ferranti Ltd.)

192

11.6. *Laser Fusion*

For many years a goal of physics has been to harness the power available in the nuclei of atoms by means of atomic fusion. The present generation of nuclear power stations are based on the principle of nuclear fission in which heavy elements are broken down to yield lighter elements and energy. The process is not as efficient as fusion and the by-products often dangerous. In fusion the idea[368] is to make atoms of light elements such as the isotopes of hydrogen, deuterium and tritium, react or ' burn ' with each other so that they fuse together to form new atomic species and a resulting massive efflux of energy. The attraction of such a technique is that the initial materials are relatively simple to obtain and the by-products are easily handled.

Previous attempts at energy generation by fusion have always tried to confine gases in a plasma and to use a magnetic field to keep the plasma away from the walls of the containing structure and to then raise the temperature sufficiently to initiate fusion. In fact a temperature of 10^8 °K is needed for deuterium/tritium fusion. This necessary condition of very high temperature has always led to failures because the magnetic fields used to confine the plasma can never be made sufficiently stable so the plasma wanders and burns out the walls of the container.

Not only is a high temperature required for the fusion process but two other interrelated conditions must be met before fusion can occur. First the density of the reacting pellet of material must be sufficiently high and secondly the particles making up the reacting material must be confined for an adequate length of time for efficient burning to take place. The generation of energy by fusion in stars relys on the huge gravitational forces to compress the reacting matter to such enormous degrees that the density is so high in stars confinement time can be very short. At the opposite extreme of thermonuclear weapons the initial fission explosion takes place so quickly that the sheer inertia of the atoms provides adequate confinement time for the fusion reaction to proceed even though the densities involved are much lower than those in the centre of a star.

The conditions of sufficient particle numbers or density and long enough reaction or confinement time can be summarized for the deuterium/tritium reaction by stating that their numerical product must exceed 10^{14}. In fusion energy generated by magnetically confined plasmas, times of 1 s and densities of 10^{14} particles are typical.

The key to the new approach to fusion generation is to rely, like the thermonuclear weapon case, on the mass inertia of the atoms to make confinement times of 10^{-12} s easily available and to increase the density to 10^{26} particles to comply with the condition outlined above. Thus the laser must do two jobs; it must heat the pellet of

Application	Advantage
Quartz and silica	Flame polished cutting of tubes and plates. No chipping. No dust. Quality of the laser cut means that no finishing processes are required.
Rubber and plastic	Non-contact cutting and drilling eliminates distortion.
Thin gauge metal	Narrow, parallel sided cuts make it superior to oxy-flame and plasma-arc. High speed profiling to fine tolerances. Minimum of finishing needed.
Nylon belting and cloth	Fast cutting speeds. Simultaneous edge sealing prevents fraying.
Fibreglass	Laser cutting eliminates costly diamond tipped tools. Smooth edged cuts. No dust.
Paper	Very fast cutting speeds. Smooth edge cut is stronger than sheared edge.

Fig. 11.9. Some materials which can be cut with the Ferranti MF400 system and the advantages gained. (Data provided by courtesy of Ferranti Ltd.).

Material	Thickness (mm)	Cutting speed (m/minute)
ABS Plastic	0·8	27·0
ABS Plastic	4·0	4·5
Acrylic	3·0	4·5
Acrylic	6·0	1·7
Asbestos	1·6	1·0
Automobile Carpet	—	18·0
Cardboard	0·1	96·0
Ceramic Tile	6·3	0·3
Engineering Felt	6·4	19·0
Formica	1·6	7·8
Glass Fibre (Resin Bonded)	1·6	5·2
Micalex	3·0	1·2
Perspex	1·6	2·4
Plywood	18·0	0·5
Silica	1·0	2·0
Sorbo Rubber	3·0	12·2
Wool Suiting	—	48·0
Galvanized Steel	1·0	4·5
High Carbon Steel	3·0	1·5
Immac 5	1·5	2·5
Mild Steel	1·0	4·5
Stainless Steel	2·8	1·2
Titanium	3·0	4·1

Fig. 11.10. Some materials, thicknesses and cutting speeds for the Ferranti MF400 system. (Data provided by courtesy of Ferranti Ltd.).

deuterium and tritium to 10^8 °K and increase the density of the pellet by a large amount. At first sight the latter requirement may seem impossible to satisfy, however, computer calculations[369] have shown that the generation of such densities is within the realms of possibility if the laser is powerful enough. A complete physical description of the processes involved would be complex but basically what happens is that when the laser beam impinges on the pellet it heats up and the hot gases evaporating from the surface at high speed react on the spherical surface by Newton's law to produce a force which compresses the pellet to one twentieth of its initial diameter, i.e. its density is increased about 10 000 times. In this way it is hoped that all the conditions necessary for laser fusion to take place will be met. The problems associated with earlier fusion schemes (such as the ZETA machine at Harwell) will be manageable because stability of the burning plasma is only required for picoseconds instead of seconds.

At the Livermore laboratory in the U.S.A. a programme to investigate the feasibility of laser fusion is underway which is probably fairly typical of several now under development throughout the world. It is proposed that a mode locked Nd–YAG laser be used to produce suitably shaped pulses of 10^{-3} J at a rate of about 30 each second. A cascade of optical amplifiers is used to increase the power of each pulse about a million times so that the final pulse emitted from the system has an energy of about 1000 J. It is ultimately envisaged that up to 20 such laser systems be built to deliver uniformly 10 000 J pulses onto a spherical pellet of deuterium and tritium about ohe millimetre in diameter. This apparatus will enable many of the practical problems to be investigated and will prove the viability of the concept although it is not envisaged that useful amounts of power will be generated. The manifestation of the latter lies further in the future but estimates have been made that the availability of a 300 kJ laser delivering 100 pulses per second will enable a 1000 MW power station to be constructed. It may not be very long before such lasers are built.

CHAPTER 12

holography

Possibly the most exciting application of the laser is in a new type of photography called holography. Unlike conventional photography in which a two-dimensional picture is produced by a lens, holography needs no image-forming lenses and the final picture is a three-dimensional view of the object or scene.

The beginnings of holography took place before the invention of the laser and the early results bear no comparison with holograms made with a laser.

Dennis Gabor, then of the British Thomson-Houston Company, made the first hologram in 1948[268,269,270] in an attempt to show how the limitations in resolution of the electron microscope might be overcome. At that time resolving powers were restricted to about 12 Å by the defects in magnetic lenses used in the instrument. It was thought that by recording both the amplitude and phase of the distorted electronic wavefront it might be possible, by subsequent reconstruction with an optical wavefront, to correct the aberrations with optical lenses. As it turned out this application was not successful and improvements in electron microscope lenses now enable resolutions of a few Angstroms to be obtained. This limit is set by other factors such as variations in magnetic fields, vibration and contamination of the object.

The essence of Gabor's work was the recording of the wavefront emanating from an object. At any point on this wavefront there will exist a certain amplitude and phase relative to other points on the wavefront. Only by recording both the amplitude and phase can *all* the information about the object be recorded. In conventional photography the photographic emulsion, being a square-law detector, records only the *intensity* of the light and not the phase so that the three-dimensional nature of the scene is lost.

The recording of phase as well as amplitude is achieved in holography by making the wavefront from the object interfere with another wavefront which is usually flat or spherical and which is known as the reference wave or reference beam. The interference fringes so produced are recorded on a conventional photographic plate which is then known as a hologram—a derivation from the Greek word ' holos ' meaning ' the whole ', i.e. the whole of the information.

To understand how the fringe pattern in a hologram records the amplitude and phase, consider again Young's two-slit experiment. Suppose one of the slits is fixed and is illuminated with light of constant

196

intensity.   Suppose also that the intensity with which the other slit is illuminated can vary, as well as its position relative to the fixed slit. The latter can be thought·of as providing the reference wavefront and the movable slit the object wavefront.   As discussed in Chapter 2 such an arrangement will form a set of parallel fringes on a screen.   If the two slits are equally illuminated, black and bright fringes will be observed and if the illumination is not equal, dark (not black) and bright fringes will be observed, i.e. the contrast of the fringes will be reduced. If on the other hand the position of the object slit is moved then the fringes will also move.   Thus information about the amplitude and the phase of the object wave is available by virtue of the contrast of the fringes and the position of the fringes respectively.   By placing a photographic plate so as to record the fringe pattern the information is permanently recorded.

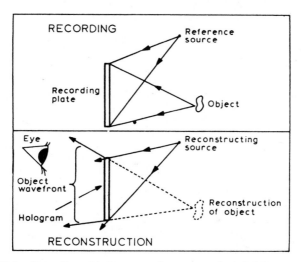

Fig. 12.1.   Recording of hologram and reconstruction of object wavefront.

In an actual holographic recording system the reference slit is replaced by a point source of light and the object slit by the illuminated object itself.   Having recorded the hologram it remains to extract the information about the object, i.e. to reconstruct the object wavefront. This process is shown in fig. 12.1 and is carried out by illuminating the hologram with the reference wavefront, both of which occupy the same positions as during the recording.   The reference wavefront, which now becomes the reconstructing wavefront, is diffracted by the hologram so that the transmitted wave is an exact replica of the original object wave.   Thus if an observer looks at this wavefront, an image of the object is seen exactly as it appeared originally.   It should now be obvious why holography is also known as wavefront reconstruction.   As the

197

reconstructed wave is an exact replica of the original it is possible, by changing the viewpoint of the observer, to examine the 'object' from different positions in exactly the same way as the actual object might have been inspected. This ability is limited only by the size of the hologram—in fact cylindrical holograms have been made in which the 'object' can be viewed from all sides. Figures 12.14 *a* and 12.14 *b* show two different views through a hologram. The three-dimensional nature of the reconstruction is evident from the change in relative positions of detail within the scene.

To gain further insight into the reconstruction process consider the interference between two plane waves as is the situation in fig. 12.2.

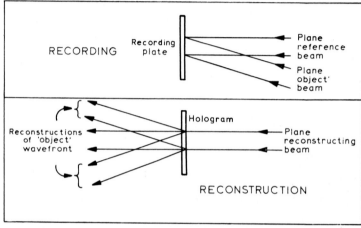

Fig. 12.2. The interference of two plane wavefronts to form a hologram which consists of straight line parallel fringes. This grating-like hologram then reconstructs the 'object' wavefront in the two first order diffracted beams.

The interference pattern consists of a set of parallel straight line fringes and can be thought of as a hologram of a plane wave using a plane wave reference beam. The resulting hologram is akin to a diffraction grating and, as is well known, if a plane wave is incident on a diffraction grating the transmitted wave consists of three collimated beams[1,2] as is also shown in fig. 12.2. The two beams on either side of the directly transmitted beam can be thought of as reconstructions of the original plane wave object beam. There are in fact *two* reconstructions of the object wavefront. The difference between these is brought out in fig. 12.3 which shows the plane wave object beam replaced by a point source. In this case there are two waves produced in the reconstruction. One wave appears to diverge from the object point and provides the primary, in this case virtual, image of the object. The second wave converges to form a second image, in this case real, known as the conjugate image. It should be noted that the primary and conjugate images can be real or virtual depending on the geometry of the holographic recording system.[271]

If the object and reference beams are incoherent, interference will not take place and a hologram cannot be made. Consequently the use of a laser as a light source greatly facilitates the making of holograms. Gabor's original holograms were made without a separate reference

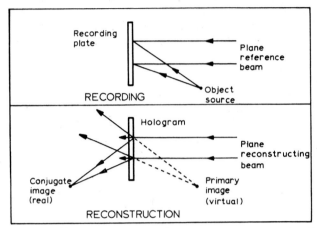

Fig. 12.3. Formation of primary and conjugate images of a point object.

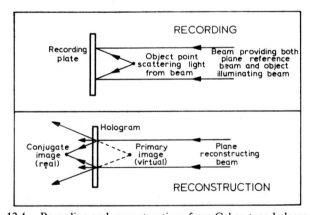

Fig. 12.4. Recording and reconstructions from Gabor-type hologram.

beam as such. The reference beam was provided by light passing, undisturbed, round the opaque object as shown in fig. 12.4.

It will be recalled that the coherence length of a light source is a measure of the discrepancy in pathlengths in an interferometer before interference fails to take place. It is obvious that the coherence length of the source is of crucial importance in holography. Gabor used high pressure mercury lamps whose coherence length was only 0·1 mm and consequently was restrained to the arrangement of fig. 12.4. This system had two major disadvantages. First, as the reference beam had to pass round the object, the latter had to take the form of small opaque regions (in fact letters) on a clear background. Solid, three-dimensional, objects could not be recorded with this system. The second disadvantage arises from the position of the primary and conjugate images which are ' in line ' and results in the primary virtual image being degraded by an out-of-focus conjugate image.

The ability to make holograms using separate object and reference beams, as indicated in fig. 12.3, became possible in 1960 with the invention of the helium-neon laser. In fact, in 1963, Leith and Upatnieks,[272] of the University of Michigan, made the first hologram of this type which is usually referred to as an ' off-axis ' hologram in contrast to Gabor's ' in line ' holograms.

12.1. *Diffuse Object Illumination*

The objects used in the first holograms of Leith and Upatnieks were transparencies and a considerable improvement was made in 1964[273] when they used diffuse illumination of the object transparency. Without a diffuser any point on the transparency is illuminated by the single reference source (which is at infinity in the case of a plane wave) and thus the information about that particular point of the object is confined to a fairly small region of the hologram. The effect of diffuse illumination is to replace the single object illuminating source with a multiplicity of sources so that information about any point of the object transparency is distributed uniformly over the entire hologram. In this way blemishes on the hologram have almost no effect on the quality of the reconstruction. Such holograms are said to be redundant while non-redundant holograms are produced by non-diffuse object illumination.

The holographic fringes formed with non-diffuse and diffuse object illumination are different in character. In the former case they appear as straight, almost parallel, lines, modulated in spacing and contrast. Diffuse illumination of the object gives the holographic fringe pattern the appearance of a random granular structure. In either case the fringe size is usually so small that it is invisible to the eye and can only be observed with a microscope.

Another advantage of diffuse object illumination is that it enables the observer to view the entire object transparency without moving the head. This is evident if a transparency is held between the eye and a small light source. Only that part of the transparency adjacent to the line between the source and the eye is visible. If a diffuser is placed behind the transparency the latter becomes visible in its entirety.

Transparencies form special cases of objects and most solid objects have a rough surface which acts as a built-in diffuser. Consequently

200

when making holograms of solid objects a separate diffuser is not required.

An interesting property of redundant holograms is that quite small regions of the hologram receive information about the complete object. In consequence a hologram of this type may be broken up and each fragment will still reconstruct to give an image of the entire object.

12.2. Speckle Pattern

If any rough surface is illuminated with a laser beam the surface appears to be covered with an irregular granular pattern[274,275] which is generally referred to as a speckle pattern. The surface of the object can be thought of as a very large number of point sources, each radiating with different amplitude and phase. As a laser is used to illuminate the surface, all the points radiate coherently with the result that they generate a random interference pattern at all points in space between the object and the observer. The speckle is therefore not localized at the surface. This is evident from the fact that it is always possible for the eye to see the speckle pattern sharply even when spectacles, if worn, are removed. Another property of the speckle pattern is that its size changes according to the resolving power of the viewing system. This can be demonstrated by viewing a speckle pattern through a small pinhole in front of the eye—the speckle appears much coarser because the resolving power of the eye has been reduced.

Speckle pattern has important consequences in holography. Any redundant hologram will form an image (real or virtual) degraded by speckle pattern. If, as is required in some applications of holography, very small holograms are made with low resolving power, the speckle pattern may be so large that the image is quite unsatisfactory.

12.3. Simple Mathematical Analysis of Holography

It will be useful to give a very simple mathematical account of how an image is formed holographically. Referring to fig. 12.5, O and R

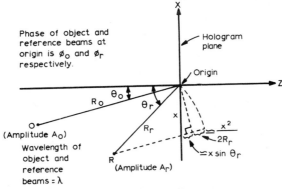

Fig. 12.5.   Holographic recording parameters.

201

are object and reference sources, distance R_o and R_r from the origin respectively. Each source is radiating coherently at a wavelength λ and the one-dimensional hologram is to be recorded in the x plane. Provided θ_o and θ_r are small, the amplitude U at some point x on the hologram is given by

$$U_o(x) \approx A_o \exp\left\{i\left[\phi_o + \frac{2\pi}{\lambda}\left(x\sin\theta_o + \frac{x^2}{2R_o}\right) + \omega t\right]\right\} \quad (12.1)$$

for the object beam, and

$$U_r(x) \approx A_r \exp\left\{i\left[\phi_r + \frac{2\pi}{\lambda}\left(x\sin\theta_r + \frac{x^2}{2R_r}\right) + \omega t\right]\right\} \quad (12.2)$$

for the reference beam and where A_o and A_r are the amplitudes of the object and reference beams respectively. ϕ_o and ϕ_r are the respective phases at the point on the hologram where $x = 0$ and $\omega = 2\pi\nu$ where ν is the frequency of the light.

As the object and reference waves are coherent the intensity $I(x)$ in the hologram plane is given by adding the amplitudes and multiplying by the complex conjugate :[1,2]

$$I(x) = (U_o + U_r)(U_o + U_r)^* \quad (12.3)$$

This intensity distribution is recorded by the photographic emulsion. In holography the relevant characteristic curve of the recording material is that of amplitude transmission v. exposure. It is assumed that a linear part of this curve is used (fig. 12.16 shows typical characteristic curves). The processed hologram is now replaced and illuminated by the reference beam only. The hologram generates a wavefront whose intensity is proportional to $I(x) \cdot U_r(x)$, i.e.

$$(A_o{}^2 + A_r{}^2)A_r \exp\left\{i\left[\phi_r + \frac{2\pi}{\lambda}\left(x\sin\theta_r + \frac{x^2}{2R_o}\right) + \omega t\right]\right\}$$

$$+ A_o A_r{}^2 \exp\left\{i\left[\phi_o + \frac{2\pi}{\lambda}\left(x\sin\theta_o + \frac{x^2}{2R_o}\right) + \omega t\right]\right\}$$

$$+ A_o A_r{}^2 \exp\left\{i\left[2\phi_r - \phi_o + \frac{2\pi}{\lambda}\left(2x\sin\theta_r - x\sin\theta_o\right.\right.\right.$$

$$\left.\left.\left. + \frac{x^2}{R_r} - \frac{x^2}{2R_o}\right) + \omega t\right]\right\} \quad (12.4)$$

The first term represents an attenuated reconstructing wave. The second represents a wave identical to the original object wave and is thus responsible for the primary, virtual, image of the object. The third term represents the conjugate, real, image. As any object can be regarded as a collection of point sources, this analysis can be extended to actual objects.

Consideration of the third term of equation 12.4 will show that the conjugate image is, in general, distorted. An undistorted real image is only obtained when a reconstructing beam is used which converges to the position originally occupied by the reference source, i.e. to the point R in fig. 12.5. In this case a real image of the object is formed at O.

An important property of a hologram is that if the *object* beam illuminates the hologram the *reference* wave will be reconstructed.

12.4. *Image Magnification*

The images formed by holograms can be magnified or diminished in three ways :

(1) By using a reconstructing wavefront having a different radius of curvature from that of the reference wavefront.

(2) By changing the size of the hologram itself.

(3) By using different wavelengths for the reference/object and reconstructing beams.

The lateral and longitudinal magnifications are given by M_{lat} and M_{long} respectively in the following equations[276,336]

$$M_{lat} = m \left(1 + \frac{m^2 z_0}{\mu z_c} - \frac{z_0}{z_r} \right)^{-1} \qquad \text{for the} \qquad (12.5)$$
$$M_{long} = \frac{1}{\mu} M^2_{lat} \qquad \qquad \begin{array}{c} \text{primary} \\ \text{image} \end{array} \qquad (12.6)$$

$$M_{lat} = m \left(1 - \frac{m^2 z_0}{\mu z_c} - \frac{z_0}{z_r} \right)^{-1} \qquad \text{for the} \qquad (12.7)$$
$$M_{long} = \frac{1}{\mu} M^2_{lat} \qquad \qquad \begin{array}{c} \text{conjugate} \\ \text{image} \end{array} \qquad (12.8)$$

where

$z_0 = z$ coordinate of the object point.
$z_c = z$ coordinate of the reconstructing point.
$z_r = z$ coordinate of the reference point.
$\mu = \lambda_c / \lambda_0$
λ_c = wavelength of the reconstructing beam.
λ_0 = wavelength used in recording.

and where m is the ratio by which the hologram is enlarged.

The following deductions can be made from equations 12.5 − 12.8 :

(1) If $z_c = \infty$, i.e. a plane reconstructing wave is used, then changing the wavelength does not result in magnification, otherwise increasing the wavelength in the reconstruction stage provides magnification.

(2) In general the lateral and longitudinal magnifications are unequal and an undistorted, unaberrated (see section 12.7), stereo image can only be obtained by making

$$z_0 = z_r = z_c, \; m = 1 \text{ and } \mu = 1.$$

203

12.5. *Fourier Transform Holography*

Holographic systems can be regarded as falling between two extreme cases: if the reference source is at infinity and the object is close to the hologram then the latter can be described in terms of a Fresnel transform and is therefore referred to as Fresnel transform hologram. Most holograms are of this type. When the object and reference source are both at infinity the hologram can be described in terms of a Fourier transform and is known as a Fourier transform hologram or a Fraunhofer hologram.

A close approximation to a Fourier transform hologram can be made[277] by placing the object and reference source close together in comparison with their distance from the hologram as in fig. 12.6.

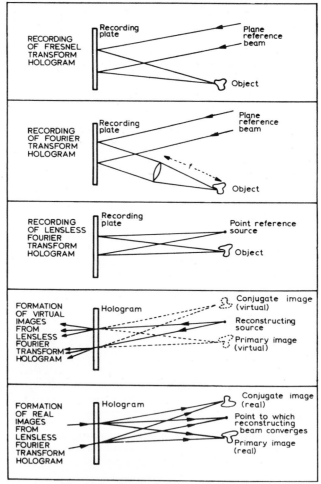

Fig. 12.6. Fourier transform holography.

The fundamental difference between Fresnel and Fourier transform holography lies in the nature of the holographic fringes. In the Fresnel case each point on the object interferes with the reference beam to form an interference pattern resembling a Fresnel zone plate of circular fringes. Fourier transform holographic fringes are almost straight and parallel. Figure 12.6 shows how two virtual images are formed from a Fourier transform hologram by using the reference source as the reconstructing source and how two real images can be formed by using a reconstructing beam which converges to the point originally occupied by the reference source. The advantage of the Fourier transform hologram will be discussed in the next section.

12.6. *Resolution in Holography*

There are four factors to be considered in discussing the resolving power of a hologram. These are

(1) The angle 2θ subtended at the object by the hologram. It is assumed that the size of the object is negligible.
(2) The wavelength λ of the light.
(3) The resolving power of the holographic recording material.[278]
(4) The size of the reference source.[279]

Factors (1) and (2) are precisely the same as those applicable to a conventional lens—the resolving power of a hologram can never be better than that of a lens of the same size and focal length (the focal length of a hologram being the distance from the hologram to the object). It should also be noted that the depth of focus of a hologram is the same as that of a lens of identical focal length and aperture.

If the smallest detail in the object is of spatial frequency $1/d_o$ where d_o can be regarded as the period of a grating of the same spatial frequency then, if the angle θ which the first order diffracted beam makes with the zero order (assuming normal incidence),

$$d_o = \frac{\lambda}{\sin \theta} \qquad (12.9)$$

Figure 12.7 *a* shows the two first order beams diffracted by the grating which, by Abbe's theory,[1] must be intercepted by a lens (or hologram recording plate) for the grating to be resolved.

In addition to falling within the aperture of the hologram, these two first order beams must also each interfere with the reference beam. The resolution of a hologram is therefore limited not only by its size but also by the ability of the recording material to record the holographic fringes.

The spacing d_h of the fringes formed by two beams interfering at an angle 2α is given by

$$d_h = \frac{\lambda}{2 \sin \alpha \cos \beta} \qquad (12.10)$$

205

where β is the angle the plane of the hologram makes with the normal to the line bisecting the angle between the two beams as indicated in fig. 12.7 b.

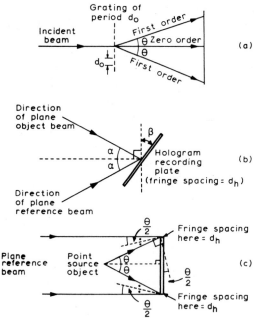

Fig. 12.7. Parameters involved in (a) resolving power of optical system of angular aperture 2θ, (b) holographic fringe spacing and (c) resolution of Fresnel transform hologram.

Suppose, as is shown in fig. 12.7 c, that the reference beam is plane and normal to the hologram and that the hologram is just wide enough to intercept the first order beams from a grating representing the maximum spatial frequency in the object. The holographic fringe spacing, which is a minimum at the extremities of the hologram, is obtained from equation 12.10 by substituting $\alpha = \theta/2$ and $\beta = \theta/2$:

$$d_h = \frac{\lambda}{2 \sin \theta/2 \cos \theta/2} \tag{12.11}$$

$$\therefore \qquad\qquad d_h = \frac{\lambda}{\sin \theta} \tag{12.12}$$

Comparing equations 12.9 and 12.12 it is apparent that

$$d_o = d_h \tag{12.13}$$

i.e. in order to record a given spatial frequency in the object the holographic recording medium must be capable of recording fringes of the same spatial frequency.

206

In the situation described above, the reference beam is obscured by the object. If, as is usually the case, this is undesirable, the reference beam must be introduced at an angle θ to the normal to the hologram. This has the effect of doubling the required resolving power of the recording medium at one extremity of the hologram and halving it at the other.

In Fourier transform holography, where the reference source is not at infinity but is located at the same distance, f, from the hologram as the object, the resolving power of the recording medium can be much less. If the separation of the object and reference point is Δ then the minimum fringe spacing, d_f can be deduced from equation 12.10, i.e.

$$d_f = \frac{2f\lambda}{\Delta} \qquad (12.14)$$

For a hemispherical recording medium, of which the dotted line represents a plane containing the reference and object sources, the fringe spacing would be smallest at P.

(a)

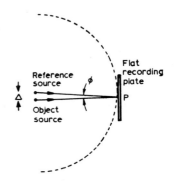

The fringes formed by interference between the dashed and dotted rays are of equal spacing if the diameter of the reference source is as large as the smallest detail in the object. The former therefore represents the resolution limit.

(b)

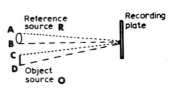

Fig. 12.8. Fringe spacing and resolution in Fourier transform holography.

The fringe spacing will not be smaller than d_f if the recording point lies on a circle whose centre is the reference source and whose radius is the distance between the reference point and the recording point as shown in fig. 12.8 a. Consequently diffraction limited resolution can always be obtained if the hologram is hemispherical and so long as the hologram recording material is able to resolve fringes of the spacing given by equation 12.14. Thus, unlike a Fresnel transform hologram, a Fourier transform hologram is able to resolve detail in the object finer than the smallest detail resolved by the recording medium.

207

The size of the reference source has a direct bearing on the resolving power of a Fourier transform hologram. Suppose that, in fig. 12.8 *b*, *R* represents a circular reference source, *A* and *B* being two ends of a diameter and *CD* represents detail in the object such that $AB = CD$. It is obvious that fringes formed as a result of interference between *A* and *C* will have the same spacing as those formed as a result of interference between *B* and *D*. It follows therefore that the distance *CD* represents the minimum resolvable detail, i.e. the resolving power of a hologram is limited to the size of the reference source. In the case of a Fresnel transform hologram the angular diameter of the source limits the angular resolution of the hologram. In both cases, of course, it is assumed that the holographic recording medium has sufficiently high resolution to record all the holographic fringes.

12.7. *Aberrations in Holography*

It can be shown[276,280,281,282] that the third order aberrations of a hologram which are spherical aberration, coma, astigmatism, field curvature and distortion can be completely eliminated by using collimated reference and reconstructing beams and ensuring that the reconstructing beam is incident from the same direction as the reference beam for reconstruction of the primary image and from the opposite direction for reconstruction of the conjugate image.

If magnification of the object imaged by a Fresnel transform hologram is required the hologram itself must also be enlarged in the same ratio as the wavelengths used in recording and reconstruction. Equation 12.5 then shows that the magnification is equal to the ratio of wavelengths.

Fourier transform holograms do not suffer from spherical aberration and the magnification produced by a change in wavelength usually does not produce sufficient aberration to justify enlargement of the hologram.

12.8. *Thick Holograms and Coloured Reconstructions*

Holographic recording materials have a finite thickness. Depending on type, silver halide photographic emulsions have thicknesses ranging between 4 μm and 15 μm. So far the thickness of the recording material has been ignored although it gives rise to interesting properties of holograms. The fringes produced by interference between two point sources *O* and *R* (regarded as object and reference sources) are three-dimensional. In fact they are hyperbolae of revolution—the axis of revolution being the line joining the points *O* and *R*.

Some of these fringes are shown in fig. 12.9 where H_1, H_2 and H_3 are three different positions of a thick hologram.

In H_1 the fringes are comparatively far apart in relation to the emulsion thickness and the hologram is effectively thin, i.e. its thickness can be disregarded.

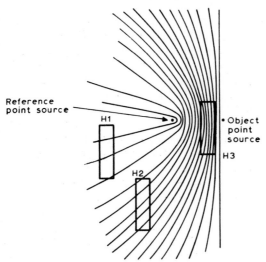

Fig. 12.9. Recording of thick holograms.

In H_2 the fringes are much closer together and the hologram can be thought of as a collection of thin holograms. A reconstruction can be made from a thin hologram, albeit greatly aberrated, whatever the position of the reconstructing source. The intensity of the image reconstructed from a thick hologram is very sensitive to the orientation of the hologram with respect to its original position in relation to the reference source. With a 15 μm thick emulsion the hologram has only to be rotated a few degrees to reduce the intensity of the image to zero. This characteristic enables many different holograms to be stored in the same photographic plate by appropriate change in the position of the reference source between each recording.[283,284,285] By moving the reconstructing source (or rotating the hologram) each reconstruction can be viewed in turn with negligible cross-talk between adjacent images. The thicker the emulsion the greater the number of holograms that can be recorded before cross-talk becomes unacceptable. Holograms made in the position H_2 are called thick transmission holograms.

If the recording plate is placed at H_3 the reference and object sources are on opposite sides of the hologram and the fringes so formed are parallel with the plane of the emulsion. From equation 12.10 the spacing of these fringes is $\lambda/2$ so as many as 50 can be recorded in a 15 μm thick emulsion. Such holograms are called reflection holograms[286] because the reconstructed object wavefront is viewed in reflection, i.e. from the same side of the hologram as the reconstructing source. Unlike thin holograms where light of any wavelength will effect a reconstruction, the intensity of the image is very sensitive to wavelength. Unless the correct wavelength is used no reconstruction can be made. This feature of reflection holograms allows reconstruction with white

209

light.[287]   The image appears coloured according to the wavelength used in the recording.   If the hologram is made with laser light of the three primary colours it is possible to obtain a colour reconstruction with a white light source.[288]

Colour holograms have also been made with thick transmission holograms by using three differently positioned reference sources to interfere with the light scattered from an object illuminated with the three primary colours.[289,290]   Each of the three differently coloured and appropriately positioned reconstructing sources reconstructs three object waves in each of the three primary colours.   As the three images are superimposed the object appears coloured.

Thin colour holograms have been made by dividing the hologram surface into three interlaced regions each recording a hologram in a primary colour.[291]   This 'spatially multiplexed' hologram forms a coloured image of the object when each reconstructing beam of different colour falls on the appropriate region of the hologram.

12.9. *The Efficiency of Holograms*

We have seen that holograms can be either of thin, thick transmission or reflection types.   Each is characterized by variations in the absorption of the recording medium.   With some materials however, such as photoresists, these variations may take the form of changes in refractive index.   These holograms are then referred to as phase holograms because the phase rather than the amplitude of the reconstructing wavefront is modified.   Exposed silver halide photographic emulsions can be converted into phase holograms by bleaching, details of which are given in section 12.10.5.

It can be seen therefore that there are in fact six quite different types of hologram.   By considering the efficiencies of diffraction gratings made by interfering two plane waves, an estimate of the upper limit of efficiency of a hologram can be made.   Figure 12.10 tabulates the efficiencies defined as a percentage of the incident reconstructing beam.[292]

Thick transmission or reflection holograms, provided they are of the phase type, can theoretically be 100% efficient although in practice efficiencies of 70% are rarely exceeded.

	Thin	Thick transmission	Reflection
Absorption	6·25	3·7	7·2
Phase	33·9	100	100

Fig. 12.10. Percentage efficiencies of the different types of holographic grating. These are theoretical figures and represent the upper limits of hologram efficiency.

12.10. *Practical Holography*

In this section some practical aspects of holography will be discussed including some details of apparatus and processing procedures.

12.10.1. *Lasers for Holography*

The ideal laser for holography gives a high power, long coherence length, uniphase output in a region of the spectrum at which the recording material is most sensitive. The choice between pulsed and c.w. operation depends on the application. For transient events or where stability of the holographic fringes is doubtful, a high power, pulsed laser is essential. However c.w. gas lasers give long coherence length, uniphase outputs and a continuous beam is very convenient in setting up the holographic system. By incorporating small apertures within the cavity, operating near threshold and using intracavity etalons, the quality of the solid state laser output can be improved. Generally speaking though, it is probably fair to say that pulsed lasers should be avoided for holography if possible.

Most gas lasers used for holography are of the helium-neon or argon types. The argon laser can typically give a watt of power in the blue-green region of the spectrum where silver halide emulsions are most sensitive. Powers available from helium-neon lasers do not exceed 100 mW, although the coherence length of 10–20 cm exceeds the few centimetres coherence length of the argon laser and they are very much cheaper.

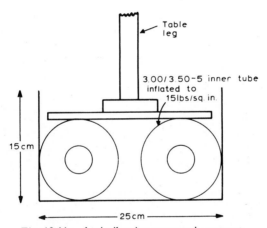

Fig. 12.11. Anti-vibration pneumatic support.

12.10.2. *Fringe Stability*

It is essential in any holographic system that during the exposure of the hologram the fringes do not move by more than about $\frac{1}{4}$ of their spacing. This implies that the holographic plate holder, the object and other components such as mirror mounts in the system must be stable. To this end these components must be attached to a rigid surface by, for example, using magnetic bases in conjunction with a steel table. The table must also be insulated from vibration of the floor in some way.

211

A cheap and effective method is to inflate small inner tubes, of the type used in go-kart wheels, inside a metal cylinder and support the legs of the table by means of a steel plate on top of the tube as indicated in fig. 12.11.

It is also desirable to suppress air currents which may arise from temperature differences. Covers placed over the table during the exposure usually eliminate this cause of fringe instability.

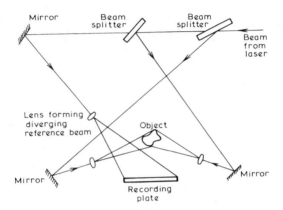

Fig. 12.12. A practical holographic recording system.

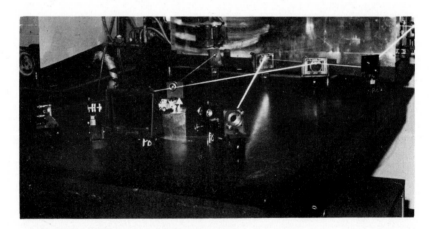

Fig. 12.13.

12.10.3. *A Typical Holographic System*

Figure 12.12 shows a practical holographic system for making holograms of solid objects. Figure 12.13 is a photograph of the system in use. All the components are mounted on the steel table with magnetic bases.

212

Fig. 12.14 (*a*).

Fig. 12.14 (*b*). Two photographs from different viewpoints of the virtual image formed by a hologram. (SERL photographs).

The object is illuminated from two directions to avoid shadows and usually a small pinhole is placed at the focus of the lens providing the diverging reference beam. This pinhole, normally about 10 μm in diameter, acts as a spatial filter and removes blemishes from the reference beam which would otherwise manifest themselves as diffraction patterns on the hologram. Figures 12.14 *a* and 12.14 *b* are two photographs taken with an ordinary camera of the virtual image of the object. The pictures were taken from slightly different viewpoints so that the three-dimensional nature of the hologram reconstruction is evident.

12.10.4. *Recording Materials*

A great variety of recording materials have been used for holography including silver halide photographic emulsions, alkali halides, photoresists,[294,295] photochromic glasses[296,297,298,299] and dichromated gelatine.[300,301,302] The first of these is by far the most popular and easy to use so the others will not be discussed here.

Emulsion Type		Approximate Resolving Power (line pairs mm⁻¹)	Approximate Energy Densities (μJ cm⁻²) for Amplitude Transmission of 0·5	
			at 4880 Å	at 6328 Å
Agfa	10E56	1000	1	not red sensitized
	14C75	1500	1	0·12
	10E75	2800	15	2
	8E75	3000	60	7·5
Kodak	649F	3000	35	35
	HR	3000	35	not red sensitized

Fig. 12.15.   Some approximate data on commonly used silver halide photographic emulsions.

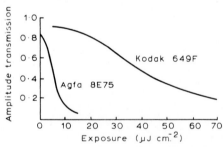

Fig. 12.16.   Typical characteristic curves for Agfa and Kodak silver halide holographic plates.   The actual curves obtained will vary according to development technique.

The most frequently used silver halide plates are tabulated in fig. 12.15 which indicates their approximate resolving power and the energy density required to produce an amplitude transmission of 0·5. Kodak H.R. and Agfa 10E56 are not sensitive to red light and are therefore unsuitable for use with helium-neon lasers.

As was mentioned earlier, the relevant characteristic curve for holographic work is the amplitude transmission v. exposure curve. Figure 12.16 shows examples of these for Agfa 10E75 and Kodak 649F; however the actual curve obtained depends greatly on development procedure.

Suppose the object wavefront is plane and is of equal intensity to the plane reference wavefront. Then the fringes will have minima of zero intensity. Reference to fig. 12.16 shows that only a certain region of the curve is linear and so some distortion will occur.[303,304,305,306] The non-linear recording gives rise to higher order diffracted beams which reduce efficiency and act as spurious images. This distortion can only be avoided by working exclusively on the linear region of the characteristic. This is accomplished by making the reference beam more intense than the object beam so that the minima of exposure of the holographic fringes are never zero. A ratio of between 4:1 and 10:1 is usually found to be satisfactory.[307]

12.10.5. *Processing of Agfa and Kodak Materials*

Two development and bleaching procedures are now given to produce high efficiency holograms. Each process is recommended by Agfa[308] and Kodak[309] for their respective materials.

Agfa

Development : 4–5 minutes in either G3p or Metinol U.

Fixing : 4 minutes in G334.

Bleach for 1 minute in a solution made up as follows :

potassium dichromate	5 g
sulphuric acid	5 ml
water	1 l

Clear for 1 minute in a solution made up as follows :

sodium sulphite	50 g
sodium hydroxide	1 g
water	1 l

Fix for 3 minutes in G334.

Kodak (recommended for 649F)

Using continuous agitation at 75°F :

Develop for 5–8 minutes in SD48.

Immerse in stop bath SBI for 15 seconds.

Rinse in running water for 1 minute.

Bleach in R9 for 2 minutes.

Wash in water for 5 minutes.

Remove stains by immersion in S 13.

Drying is critical. To ensure uniform drying :

Rinse in methanol diluted 1–7 with water,

then wash twice in iso-propyl alcohol.

The solutions should be made up as follows : (N.B. some of these chemicals are hazardous)

215

SD48

solution A :	water	750 ml
	dessicated sodium sulphite	8 g
	pyrocatechol	40 g
	dessicated sodium sulphate	100 g
	add water to make	1 l
solution B :	water	750 ml
	sodium hydroxide	20 g
	dessicated sodium sulphate	100 g
	add water to make	1 l

Mix equal parts of A and B immediately before use. The individual solutions have a life of one month unmixed (in stoppered bottles) and a life of 15 minutes mixed.

SBI

| water | 1 l |
| acetic acid | 17 ml of 80% solution |

R9

water	1 l
potassium dichromate	9·5 g
conc. sulphuric acid	12 ml

S13

solution A :	water	750 ml
	potassium permanganate	2·5 g
	conc. sulphuric acid	8 ml
	add water to make	1 l
solution B :	water	750 ml
	sodium bisulphite	10 g
	add water to make	1 l·

solutions A and B should be used as follows :

Bathe in A for 1 minute.

Rinse in B for 1 minute. This makes the plate less sensitive to print-out.

Discard the solutions after use.

12.11. *Applications of Holography*

The unique properties of holograms have given rise to numerous potential applications and a great deal of work is underway to overcome the practical problems involved in using holographic techniques. Only a brief outline of some of the uses of holography is given, not only for reasons of space but also because the field is in a rapid state of flux and some applications are more interesting or practicable than others.

12.11.1. *Holographic Interferometry*

All interferometers work by comparing the shape of two or more wavefronts. The wavefronts so conveniently generated by holograms immediately offer several methods of interferometry which can be divided into four main types : (*a*) live fringe interferometry, (*b*) frozen fringe interferometry, (*c*) time averaged interferometry and (*d*) contour generation. These will now be described in turn.

Fig. 12.17. Comparison of cylinder bores by holographic interferometry. The fringes indicate that the axis of the test cylinder is curved in comparison with the master cylinder. (NPL photograph).

(*a*) Suppose a hologram is made and, after processing, is carefully replaced in its original position with respect to the object and reference/reconstructing beam. The virtual image so produced is superimposed on the original object. If the object continues to be illuminated with laser light, any deformation it may suffer subsequent to the making of the hologram will be revealed as interference fringes localized on or near the object.[310,311,312,313,314,315] The position and number of the fringes may be changed by changing the force acting on the object. Such a technique

217

can give valuable information about three-dimensional objects with hitherto unimaginable ease.

For engineering purposes holographic interferometry is rather too sensitive. This is because the surface roughness of most engineering components is more than the wavelength of light so overall changes in the mean position of the surface are lost. One way of overcoming this problem is to use very oblique illumination of the surface under test which then acts as a mirror and surface roughness effects are eliminated. Figure 12.17 shows the reconstruction of an obliquely illuminated car engine cylinder bore looking down the bore. Each fringe is equivalent to a change in height of 1·8 μm and a slight curvature of the test cylinder is revealed.[316]

Oblique object illumination would not be necessary if some other method could be found for desensitizing the process such as using a carbon dioxide laser whose output is at 10·6 μm wavelength as the light source. It might then be possible to compare production line objects with the virtual image from the hologram of a master object. This would achieve a tremendous saving in checking time.

The 'live' fringe method however suffers from the disadvantage that accurate replacement of the hologram after processing is very difficult, so much so in fact that it may be found easier to process the hologram in situ. Even this is not always satisfactory as some emulsion shrinkage inevitably takes place with the result that even with no object deformation a few fringes will appear.

(b) By making two holograms on the same plate before and after the object is deformed the effects of emulsion instability can be avoided as each hologram is affected identically. Real time analysis is, of course, not possible with this 'frozen fringe' technique and a sequence of such double exposure holograms must be made to study a changing situation.

Figure 12.18 shows the reconstruction from a double exposure hologram of a brass tube in a 4-jaw chuck. One of the jaws was tightened between exposures.

(c) Holograms of vibrating objects can be thought of as an infinite number of superimposed holograms. Such a hologram is known as a time-averaged hologram and gives an image of the object upon which a fringe system is superimposed which gives information about the mode of vibration of the object.[317,318,319] Suppose that a surface is vibrating with amplitude d in a direction normal to its plane and is illuminated at an angle θ with respect to the normal to the surface. Then if a hologram is recorded of the scattered light at an angle θ to the other side of the normal, fringes will be obtained whose intensity distribution is given by

$$I = I_o J_o^2 \left(\frac{4 \pi d \cos \theta}{\lambda} \right) \tag{12.14}$$

where I_o is the intensity when the object is stationary and λ is the

Fig. 12.18. Holographic interferometry of a brass tube using the 'frozen fringe' technique. One jaw of the 4-jaw chuck has been tightened between exposures. (NPL Photograph).

wavelength of the light. $\mathcal{J}_o$ indicates a Bessel function of zero order. Maxima of intensity occur when

$$d \cos \theta = \frac{n\lambda}{4} \text{ where } n = 0, 1, 2, 3 \ldots \text{ etc.} \tag{12.15}$$

When $n = 0$ there is no vibration, i.e. a node exists and the fringe will be brightest. The intensity of the maxima for other values of n decrease as n increases.

Figure 12.19 is a photograph of a reconstruction from a hologram made of a steam turbine blade clamped at its lower end and vibrating at 12,347 Hz. On a good reconstruction as many as 20 fringes may be counted thus enabling comparatively large amplitudes of vibration to be measured.

Changing the frequency of the vibrating object often changes the mode pattern as is shown in fig. 12.20 where a clamped guitar is being vibrated at the different frequencies shown.

In order to examine a continuous range of frequencies a different method can be used in which the virtual image of the stationary object is superimposed on the actual object which can be vibrated at any

Fig. 12.19. Time-averaged fringes on a vibrating steam turbine blade. (NPL photograph).

220

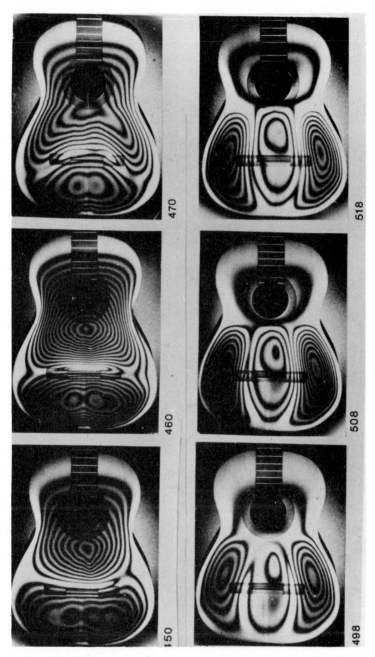

Fig. 12.20.   Time-averaged fringes on a clamped guitar vibrating at the frequencies indicated.   (Photograph from the work of K. A. Stetson at the Institute of Optical Research, Stockholm).

221

desired frequency.   If the object illumination is strobed at the *frequency of vibration* the shape of the surface at that particular frequency is revealed by the fringe pattern.   By scanning the vibration frequency which is always kept identical to the strobe frequency, the resonances of the object can be determined.   Figure 12.21 shows 17·5 cm diameter aluminium disc vibrating at 3074 Hz on which the fringes have been stroboscopically arrested.[320,321]

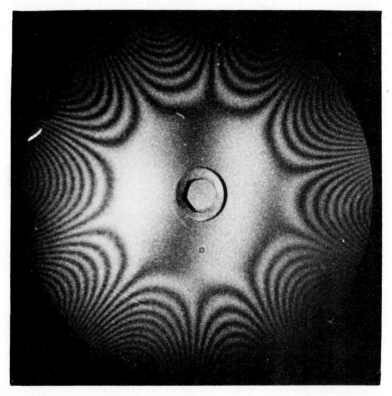

Fig. 12.21.   Stroboscopically arrested fringes on a circular aluminium plate vibrating at 3074 KHz.

(*d*) A set of contours can be generated over the surface of a ' reconstructed ' object by making a hologram with two different wavelengths λ_1 and λ_2 simultaneously.[322,323,324]   When a reconstructing beam of wavelength, say, λ_1 is used two object wavefronts are produced, one giving an image of unit magnification and another of some slightly different magnification.   The two different wavefronts then interfere to generate contours whose height interval h is given by

$$h = \frac{\lambda_1{}^2}{\lambda_1 \sim \lambda_2} \qquad (12.16)$$

Finally mention should be made of two commercial applications of holographic interferometry. Motor tyres have been tested non-destructively[325] for separations between component layers and structural peculiarities and honeycomb structures used in aircraft construction have been examined,[325] also non-destructively, for separation between the skin and the core of the honeycomb.

12.11.2. *Particle Analysis*

Holography has been used to determine the size and distribution of particles in aerosols and fogs.[326,327,328,329] Holography has the advantage over other methods of not disturbing the sample and not being limited in depth of field as would be the case with conventional photography. The method used is similar to Gabor's in-line holography system except that the reference beam interferes with the far-field or Fraunhofer diffraction pattern of the particles. Figure 12.22 shows the arrangement of the components.

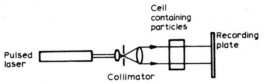

Fig. 12.22. Arrangement for recording a Fraunhofer hologram of a particle distribution.

A pulsed ruby laser is used as the light source in order to record dynamic distributions. By recording with the Fraunhofer pattern the out-of-focus conjugate image is not troublesome.

If the diameter of the particles is d and the distance between the hologram and the particle is z then, for wavelength λ, the approximate condition for recording the Fraunhofer hologram is

$$d^2/\lambda < z \qquad (12.17)$$

Reconstruction is usually observed by a fixed television camera. The hologram is then moved so that the entire volume of the sample can be examined at high magnification. Using this technique distributions containing particles down to less than 1 μm in diameter have been analysed.

Another possible application of this technique is for recording bubble chamber tracks.[330,331,332] The large chambers coming into use make conventional photography very difficult owing to the incompatibility of large depths of field with high resolution. By making a hologram, the entire volume of the chamber could be examined at leisure with high resolution.

223

12.11.3. *Character Recognition*

The complicated wavefront from an object is generated from a hologram by the simple wavefront of the reference beam. This process is reversible so that the reference wave can be generated by the object wave. This principle forms the basis of holographic pattern recognition whereby, for example, fingerprints, postal addresses and targets on aerial reconnaissance film might be recognized.[333,334]

To explain how this process might work in practice, consider fingerprint recognition where a scene-of-crime fingerprint must be matched with one contained in a library of identified prints. At the moment, despite some classification being possible, this is exceedingly laborious. With a holographic technique the library would consist of holograms of the fingerprints. By allowing the diffracted wave from a transparency of the scene-of-crime fingerprint to fall on each of the holograms contained in the library, quick identification can be made by electronic detection of the regenerated reference beam. With suitable mechanical design the entire hologram library could be quickly and automatically scanned. Unfortunately such a system is not yet in use due to the poor quality of scene-of-crime fingerprints which result in ambiguity of identification. This ambiguity is the chief problem in any holographic character recognition system especially when handwritten characters such as postal codes are concerned and it seems that it may be some time before this problem is overcome.

12.11.4. *Holographic Microscopy*

It has been explained how the image produced by a hologram can be made larger than the original object by a change of wavelength and wavefront curvature in reconstruction. Equations 12.5–12.8 summarize how the various parameters affect magnification. Gabor's original application of holography was to reconstruct holograms made with electron beams with visible light so obtaining an enormous magnification. However advances in electron microscopy have obviated the need for such a system and more mundane, although nevertheless useful, applications of holography to microscopy have been developed.

Holographic microscopy can be divided into two types according to whether the hologram provides the magnification or is used in conjunction with a conventional microscope.

If a plane reference source and a point reconstructing source near the hologram is used magnification of up to a hundred or so can be obtained. Figure 12.23 is a photograph of the magnified image of a fly's wing obtained by this means.[335,336] Images so produced are aberrated. Aberrations can only be avoided by using a longer wavelength in the reconstruction stage and enlarging the hologram in the ratio of the wavelengths. Enlargement of the hologram is not easy as the enlarging lens must resolve the holographic fringes which extend over a considerable

Fig. 12.23.   Holographic magnification obtained by using different radii of curvature for the reference and reconstructing beams.   (From reference 335).

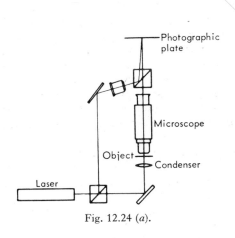

Fig. 12.24 (a).

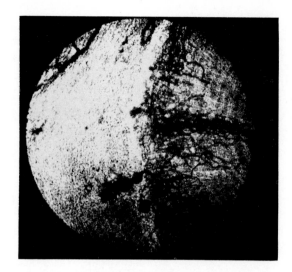

Fig. 12.24 (*b*).

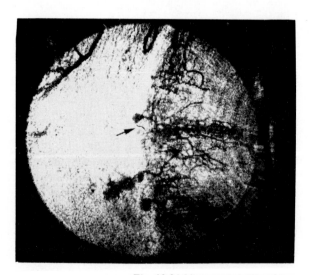

Fig. 12.24 (*c*).

Fig. 12.24 (*a*).　The optical arrangement for making a hologram of a microscopic image.
Figs. 12.24 (*b*) and (*c*) are reconstructions from the hologram.　Figure 12.24 (*c*)
was obtained by defocussing the camera by 40 μm in comparison with 12.24 (*b*).
The arrow indicates detail 1 μm in size.　(From reference 337).

area. If a short wavelength laser were to be developed, direct holographic microscopy would be feasible. For instance, if a hologram were made with an X-ray laser having an output at 5 Å and the reconstruction made with visible light an immediate magnification of $1000 \times$ would be obtained. If this image were viewed with a conventional microscope a total magnification of one million would be possible.

The most immediately useful application in this field is not in obtaining magnification directly but in storing samples of material for subsequent conventional microscopy. This is especially useful in studying rapidly changing situations where a pulsed laser can be used to 'freeze' a volume of material which can then be studied at leisure. The ability of a hologram to record a volume also enables different planes in depth to be examined. The necessity for preparing slides, which often distort the specimen, is eliminated. Interferometry may also be carried out as can all the conventional microscope techniques such as dark field and phase contrast.

Two techniques have been used. First a unit magnification hologram can be made and the image examined with a microscope. For high numerical aperture recording the recording plate must be placed close to the object which may lead to difficulty in introducing the reference beam. This can be overcome by making the reference beam incident from the back of the recording plate. In either case the hologram substrate must be of the highest quality and the recording material must not distort. These requirements can be avoided by using a second technique in which the hologram is recorded *after* the image has been magnified by an ordinary microscope in the manner indicated in fig. 12.24 a.[337,338] Figures 12.24 b and 12.24 c show how different planes can be examined, in this case the separation in depth being $40 \mu m$. Detail down to $1 \mu m$ (as indicated by the arrow) can be resolved in these pictures.

12.11.5. *High Resolution Holography*

Holograms have great potential as high resolution imaging devices. Holograms of objects at unit magnification do not suffer from any aberrations, thus in certain circumstances where high resolution over a large area is required, holograms can form images which it would otherwise be impossible to form with even the best lenses. As mentioned in the previous section the hologram material must not deform in processing otherwise image quality will suffer. The gelatine present in silver halide photographic emulsions takes up water from the developing solutions and usually distorts considerably. Consequently nongelatinous recording materials, such as photoresist,[295] have been employed as recording media to get round this problem.

One attractive application of high resolution holographic imaging is in microcircuit production to avoid contact printing which damages the delicate photographic masks used to delineate the circuits.[339,340,341,342]

227

Projection of large mask patterns with a lens is impossible as the aberrations of the lens can never be sufficiently well corrected. Unfortunately this application awaits more powerful lasers whose outputs are matched to the photoresist which is coated on the silicon slices in which the circuits are made. At present these slices are also insufficiently flat to allow for the very small depths of focus at high resolutions.

12.11.6. *Holographic Diffraction Gratings*

The simplest possible hologram is the pattern formed by interference between two plane waves. The fringe pattern consists of straight parallel fringes whose regularity of spacing, provided the interfering wavefronts are sufficiently flat, may be as good as, or better, than the lines on diffraction gratings made with ruling engines.[343] These conventional diffraction gratings often suffer from ghost images as a result of periodic errors in the ruling engine. These ghosts may be particularly annoying in applications such as Raman spectroscopy and therefore there is a distinct possibility that in the future holographic gratings may compete successfully with conventional gratings for some applications.[295,344,345]

12.11.7. *Acoustic Holography*

Holograms can be made with radiation of wavelength longer than that of light and a considerable amount of work has been carried out on making holograms with microwaves and acoustic waves.[346] In either case visible light is used in the reconstruction to enable direct viewing with the eye. Only acoustic holography will be discussed as it has several important possible applications. These arise from the transmission properties of acoustic waves through solid bodies. Usually, at low frequencies, little attenuation takes place in transmission through continuous solids, most takes place by reflection at interfaces. These interfaces are important in, for example, medical diagnosis, metallurgy and in underwater and underground environments. Acoustic holography could be applied in each of these situations to provide a three-dimensional internal picture otherwise unobtainable.

Acoustic holograms can be made in exactly the same way as ordinary holograms in that an acoustic source, such as a piezo-electric transducer, provides a reference beam and an object illuminating beam. The transducer emits coherent acoustic waves and unlike visible light holography the difficulty lies in recording the acoustic interference pattern, not in obtaining a coherent source. The resulting acoustic hologram, if viewed directly with a visible light reconstructing beam, would give a greatly diminished image, perhaps $1000 \times$ reduced. The image can be enlarged without aberrations by reducing the size of the hologram in the ratio of the wavelengths although in practice this would result in holograms less than a millimetre in size. A compromise is usually made by reducing the hologram to a few centimetres in size and viewing the image with a low power microscope. Reference to equation 12.5 will show that the longitudinal magnification is now greater than unity, consequently acoustic holograms usually give reconstructions which appear two-dimensional.

One of the first methods of recording acoustic holograms used the surface of a liquid as the recording medium.[347] Figure 12.25 shows how the acoustic reference and object waves, obtained from a 7 MHz transducer, interfere to produce variations in surface height proportional to the amplitude of the wave at any particular point. The surface may then be photographed using oblique incoherent illumination —the peaks of the waves scattering more and therefore appearing bright. The resulting picture, i.e. the hologram, is then reduced in size and the reconstruction obtained by illumination with visible laser light. Alternatively the surface may be directly illuminated with a laser beam and the reconstructed virtual image viewed beneath the surface of the liquid. In this way the advantage of real time holography is gained.

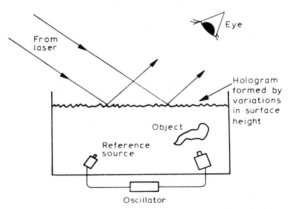

Fig. 12.25. Arrangement for making an acoustic hologram and observing the real-time reconstruction.

Much lower acoustic powers are sufficient if a microphone is used in air to scan the acoustic field.[348,349,350,351] Alternatively it has been demonstrated that the same hologram can be recorded by using a stationary detector and scanning the source.[352] Both these methods are shown in fig. 12.26 which also indicates the size of the apparatus.

The electrical output from the detector microphone is fed to a cathode ray oscilloscope in front of which is placed a photographic plate. The scanning microphone produces a spot on the tube which scans in synchronism and whose intensity is proportional to the acoustic field intensity. By scanning the entire field a complete hologram is recorded on the photographic plate. Figure 12.27 is a photograph of such a

229

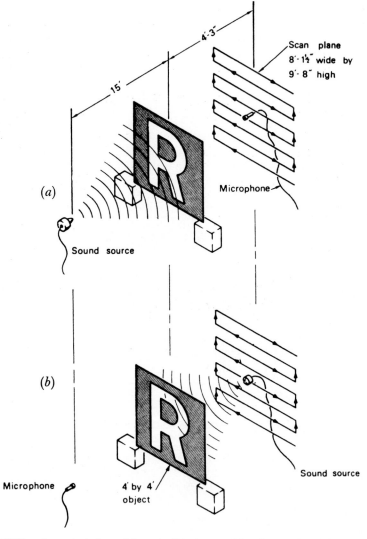

Fig. 12.26. Acoustic holographic recording system (a) using a scanning detector; (b) using a scanning source. (From reference 352).

hologram and its reconstruction. The hologram was made using the arrangement shown in the upper diagram of fig. 12.26. The great disadvantage of scanning systems is the long time taken to record the hologram.

An important feature of acoustic detectors is that, unlike photographic plates, they are sensitive to amplitude rather than intensity and so preserve phase information. This enables artificial reference beams to be supplied electronically directly to the detector so simplifying the holographic system by eliminating the need for a separate reference beam.

Fig. 12.27. An acoustic hologram and the reconstruction obtained using the apparatus shown in fig. 12.26 (a). (From reference 352).

A recent method of recording acoustic holograms involves making, with a pulsed visible light laser, two holograms on the same recording plate in rapid succession of a surface upon which the acoustic holographic interference pattern impinges. If the two pulses are separated by half the reciprocal of the acoustic frequency and the phase of the reference beam of the second hologram is changed by π with respect to that of the first, then, for a suitable acoustic amplitude, the resultant holographic interferogram will also be the acoustic hologram. This technique, although not possible in real time, enables acoustic holograms to be made in 10^{-6}s.

12.11.8 Data Storage by Holography

Future data stores for computers will need to store 10^8 bits of information with access times of less than a microsecond.[353] Although the storage capacity of magnetic tapes and discs is unlimited, access times are restricted to 10–100 ms and they are relatively bulky. Ferrite core memories can achieve the performance required but at prohibitive cost.

231

For these reasons[354,355] several research groups are investigating a completely different and novel method of computer data storage which makes use of holography.[356,357,358,359]

Figure 12.28 shows the basic components of a holographic data storage system.[356,357] The beam from the laser is deflected, usually acoustically or electro-optically, so as to address one of a large number of holograms. Systems built to date have used as many as 1024 holograms each about 1 millimetre in diameter in a 32 × 32 array. The output from each hologram consists of differing sets of focussed points of light which can be detected by an array of photodiodes whose output then conveys the appropriate instruction to the computer. A 1 millimetre diameter hologram can store 10^4 bits of information with an adequate signal-to-noise ratio. Consequently a small plate only several centimetres square can store millions of bits of information.

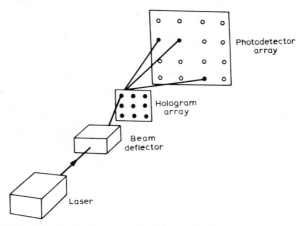

Fig. 12.28. The basic elements of a holographic data storage system.

The system described could store an equal amount of information in the form of conventional microimages but access times would be unacceptably long because of mechanical problems in quickly moving a lens to focus an image of the appropriate set of points. The self-focussing property of holograms gets round this difficulty. Another advantage of the holographic approach is that each bit of information is stored over the entire area of the hologram. This built-in redundancy greatly reduces the effect of dust or blemishes on system performance.

The array of holograms is made using conventional holographic techniques and the real images are obtained by reversing the hologram plate so as to effectively bring the reconstructing beam in from the opposite direction of the reference beam. New holograms must be made if the data needs to be updated so the system described is limited to data which does not require updating too often. Rapid insertion of new

232

holograms in situ is, however, being investigated using photochromic materials—effectively self-developing media from which any image can be erased by sufficiently intense incident light. A great deal of effort will be required however before such read in/read out systems become practicable.

REFERENCES

(1) R. W. DITCHBURN : *Light*, Blackie.
(2) F. A. JENKINS and H. E. WHITE : *Fundamentals of Optics*, McGraw-Hill.
(3) A. EINSTEIN, ' On the Quantum Theory of Radiation ', *Phys. Z.*, **18**, 6, (March 1917), pp. 121–128.
(4) J. WEBER : ' Amplification of Microwave Radiation by Substances not in Thermal Equilibrium ', *Trans. IRE, PGED-3*, (June 1953), pp. 1–4.
(5) N. G. BASOV and A. M. PROKHOROV : ' Possible Methods of obtaining Active Molecules for a Molecular Oscillator ', (in Russian), *Zh. Ebsperim. i Teor, Fiz.*, **28**, 249, (1955) (Translation : *Soviet Phys. JETP*, **1**, 184, (1955)).
(6) E. M. PURCELL and R. V. POUND : ' A Nuclear Spin System at Negative Temperature ', *Phys. Rev.*, **81**, 2, (Jan. 1951) pp. 279–280.
(7) J. P. GORDON, H. J. ZEIGER and C. H. TOWNES : ' Molecular Microwave Oscillator and New Hyperfine Structure in the Microwave Spectrum of NH_3 ', *Phys. Rev.*, **95**, 1, (1954) pp. 282–284.
(8) J. P. GORDON, H. J. ZEIGER and C. H. TOWNES : ' The Maser—New Type of Microwave Amplifier, Frequency Standard and Spectrometer ', *Phys. Rev.*, **99, 4**, (Aug. 1955) pp. 1264–1274.
(9) N. BLOEMBERGEN : ' Proposal for a New Type of Solid State Maser ', *Phys. Rev.*, **104**, 2, (Oct. 15, 1956) pp. 324–327.
(10) H. E. D. SCOVIL, G. FEHER and H. SEIDEL : ' Operation of a Solid State Maser ', *Phys. Rev.*, **105**, 2 (Jan. 15, 1957), pp. 762–763.
(11) A. L. McWHORTER and J. W. MEYER : ' Solid State Maser Amplifier ', *Phys. Rev.*, **109**, 2, (Jan. 15, 1958), pp. 312–318.
(12) G. MAKHOV, C. KIKUCHI, J. LAMBE and R. W. TERHUNE : ' Maser Action in Ruby ', *Phys. Rev.*, **109**, 4, (Feb. 15, 1958), pp. 1399–1400.
(13) A. L. SCHAWLOW and C. H. TOWNES : ' Infrared and Optical Masers ', *Phys. Rev.*, **112**, 6, (Dec. 15, 1958), pp. 1940–1949.
(14) T. H. MAIMAN : ' Stimulated Optical Radiation in Ruby ', *Nature*, **187**, 4736, (Aug. 6, 1960), pp. 493–494.
(15) D. F. NELSON and W. S. BOYLE : ' A Continuously Operating Ruby Laser ', *Appl. Opt.*, **1**, 2, 1962, pp. 181–183.
(16) A. JAVAN, W. R. BENNETT, Jr. and D. R. HERRIOTT : ' Population Inversion and Continuous Optical Maser Oscillation in a Gas Discharge Containing a He-Ne Mixture ', *Phys. Rev. Lett.*, **6**, 3, (Feb. 1, 1961), pp. 106–110.
(17) M. BORN and E. WOLF : ' Principles of Optics ', Pergamon Press, p. 318.
(18) M. BORN and E. WOLF : ' Principles of Optics ', Pergamon Press, p. 511.
(19) R. HANBURY-BROWN and R. Q. TWISS : *Phil. Mag.*, **7**, 45, (1954) p. 663.
(20) W. P. BARR : ' The production of low scattering dielectric mirrors using rotating vane particle filtration ', *Jour. Sci. Inst.*, **2**, 2, (1969) pp. 1112–1114.
(21) A. G. FOX and T. LI : ' Resonant Modes in a Maser Interferometer ', *Bell System Tech. Jour.*, **40**, (March 1961), pp. 453–488.
(22) G. D. BOYD and J. P. GORDON : ' Confocal Multimode Resonator for Millimetre through Optical Wavelength Masers '. *Bell System Tech. Jour.*, **40**, (March 1961), pp. 489–508.

(23) G. D. BOYD and H. KOGELNIK : 'Generalized Confocal Resonator Theory', *Bell System Tech. Jour.*, **41**, (July 1962), pp. 1347–1371.

(24) P. W. SMITH : 'Stabilized, Single-Frequency Output from a Long Laser Cavity', *IEEE Jour. QE-1*, **8**, (1965) pp. 343–348.

(25) D. A. KLEINMAN and P. P. KISLIUK : 'Discrimination Against Unwanted Orders in a Fabry-Perot Resonator', *Bell System Tech. Jour.*, **41**, (March 1962), pp. 453–462.

(26) H. KOGELNIK and C. K. N. PATEL : 'Mode Suppression and Single Frequency Operation in Gaseous Optical Masers', *Proc. IRE*, **50**, 11, (Nov. 1962), pp. 2365–2366.

(27) A. A. VUYLSTEKE : 'Theory of Laser Regeneration Switching', *Jour. Appl. Phys.*, **34**, (June 1963), pp. 1615–1622.

(28) R. DALY and S. D. SIMS : 'An Improved Method of Mechanical Q-Switching Using Total Internal Reflection', *Appl. Opt.*, **3**, (Sept. 1964), pp. 1063–1066.

(29) F. J. McCLUNG and R. W. HELLWARTH : 'Giant Optical Pulsations from Ruby', *Jour. Appl. Phys.*, **33**, (1962), pp. 828–829.

(30) P. KALAFAS, J. I. MASTERS and E. M. E. MURRAY : 'Photosensitive Liquid Used as a Non-Destructive Passive Q-Switch in a Ruby Laser', *Jour. Appl. Phys.*, **35**, 8, (1964), p. 2349.

(31) B. H. SOFFER : 'Giant Pulse Operation by a Passive Reversibly Bleachable Absorber', *Jour. Appl. Phys.*, **35**, 8, (1964), p. 2551.

(32) J. I. MASTERS, J. H. WARD and E. HARTOUNI : 'Laser Q-Spoiling Using an Exploding Film', *Rev. Sci. Inst.*, **34**, 4, (1963), pp. 365–367.

(33) D. G. GRANT : 'A Technique for Obtaining Single High Peak Power Pulses from a Ruby Laser', *Proc. IRE*, **51**, (1963), p. 604.

(34) E. SCHIEL and J. J. BOLMARCICH : 'Direct Modulation of a He-Ne Gas Laser', *Proc. IEEE*, **51**, 6, (1963), pp. 940–941.

(35) L. F. JOHNSON and D. KAHNG : 'Piezoelectric Optical-Maser Modulator', *Jour. Appl. Phys.*, **33**, 12, (1962), pp. 3440–3443.

(36) C. F. QUATE, C. D. W. WILKINGSON and D. K. WINSLOW : 'Interaction of Light and Microwave Sound', *Proc. IEEE*, **53**, (1965).

(37) P. DEBYE and F. W. SEARS : *Proc. Nat. Acad. Sci.*, **18**, (1932) p. 409.

(38) C. V. RAMAN and N. S. NATH : *Proc. Indian Acad. Sci.*, A2, (1935), pp. 406–412.

(39) C. V. RAMAN and N. S. NATH : *Proc. Indian Acad. Sci.*, A3, (1936), pp. 75–84.

(40) L. K. ANDERSON : 'Microwave Modulation of Light', Microwaves, 8, (Jan. 1965), pp. 42–51.

(41) H. Z. CUMMINS and N. KNABLE : 'Single Sideband Modulation of Coherent Light by Bragg Reflection from Acoustical Waves', *Proc. IEEE*, **51**, (Sept. 1963), p. 1246.

(42) K. GÜRS and R. MÜLLER : 'Wide-band Modulation by Controlling the Emission of an Optical Maser (Decoupling Modulation)', *Phys. Lett.*, **5**, 3, (July 1963), pp. 179–181.

(43) K. M. JOHNSON : 'Microwave Light Modulation by the Pockel Effect', *Microwave Jour.*, **7**, (Aug. 1964), pp. 51–56.

(44) D. HOLT : 'Laser beam deflection techniques', *Optics Tech.*, **2**, 1, (Feb. 1970), pp. 1–7.

(45) W. KULCKE, K. KOSANKE, E. MAX, M. A. HABEGGER, T. J. HARRIS and H. FLEISHER : 'Digital Light Deflectors', *Proc. IEEE*, **54**, 10, (1966).

(46) A. KORPEL, R. ADLER, P. DESMARES and W. WATSON : 'A Television Display Using Acoustic Deflection and Modulation of Coherent Light', *Proc. IEEE*, 54, (1966).

(47) L. K. ANDERSON, S. BROJDO, J. T. LAMACCHIA and L. H. LIN : ' A High Capacity Semi-Permanent Optical Memory ', *IEEE Jour. Quan. Elec.*, **3**, 6, (1967).

(48) C. E. BAKER : ' Laser Display Technology ', *IEEE Spectrum*, **5**, (1968).

(49) P. A. FRANKEN, A. E. HILL, C. W. PETERS and G. WEINREICH : ' Generation of Optical Harmonics ', *Phys. Rev. Lett.*, **7**, 4, (Aug. 15, 1961), pp. 118–119.

(50) P. D. MAKER, R. W. TERHUNE, M. NISENOFF and C. M. SAVAGE : ' Effects of Dispersion and Focusing on the Production of Optical Harmonics', *Phys. Rev. Lett.*, **8**, 1, (Jan. 1, 1962), pp. 21–22.

(51) J. A. GIORDMANE, ' Mixing of Light Beams in Crystals ', *Phys. Rev. Lett.*, **8**, 1, (Jan. 1, 1962), pp. 19–20.

(52) R. G. SMITH, K. NASSAU and M. F. GALVIN : ' Efficient Continuous Optical Second-Harmonic Generation ', *Appl. Phys. Lett.*, **7**, 10, (Nov. 15, 1965), pp. 256–258.

(53) J. E. GEUSIC, H. J. LEVINSTEIN, S. SINGH, R. G. SMITH and L. G. VAN UITERT : ' Continuous $0.532 \mu m$ Solid-State Source Using $Ba_2 Na Nb_5 O_{15}$ *App. Phys. Lett.*, **12**, 9, (May 1, 1968), pp. 306–308.

(54) W. R. C. ROWLEY and D. C. WILSON : ' Wavelength Stabilization of an Optical Maser ', *Nature*, **200**, 4908, (1963), pp. 745–747.

(55) R. A. McFARLANE, W. R. BENNETT, JR., and W. E. LAMB, JR. : ' Single Mode Tuning Dip in the Power Output of a He–Ne Optical Maser, *Appl. Phys. Lett.*, **2**, 10, (May 15, 1963), pp. 189–190.

(56) American Institute of Physics Handbook, McGraw Hill, 1957 and D. S. McCLURE : ' Electronic Spectra of Molecules and Ions in Crystals ', Ed. F. Seitz and D. Turnbull, Academic Press, pp. 400–459.

(57) D. ROSS : *Lasers, Light Amplifiers and Oscillators*, Academic Press, (1969), pp. 421–455.

(58) T. H. MAIMAN, R. H. HOSKINS, I. J. D'HAENENS, C. K. ASAWA and V. EVTUHOV : ' Stimulated Optical Emission in Fluorescent Solids. II. Spectroscopy and Stimulated Emission in Ruby ', *Phys. Rev.*, **132**, 4, (1961), pp. 1151–1157.

(59) T. H. MAIMAN : ' Optical and Microwave—Optical Experiments in Ruby ', *Phys. Rev. Lett.*, **4**, 11, (June 1, 1960), pp. 564–566.

(60) G. MAGYAR : ' Mode Selection Techniques for Solid-State Lasers ', *Optics Tech.*, **1**, 5, (Nov. 1969), pp. 231–239.

(61) M. HERCHER : ' Single-Mode Operation of a Q-Switched Ruby Laser ', *Appl. Phys. Lett.*, **7**, 2, (July 15, 1965), pp. 39–41.

(62) L. F. JOHNSON : ' Optical Maser Characteristics of Rare-Earth Ions in Crystals ', *Jour. Appl. Phys.*, **34**, 4, (1963), pp. 897–909.

(63) E. SNITZER : ' Optical Maser Action of Nd^{3+} in a Barium Crown Glass ', *Phys. Rev. Lett.*, **7**, 12, (Dec. 1961), pp. 444–446.

(64) E. SNITZER : ' Neodymium Glass Laser ', Quantum Electronics II. Ed. P. Grivet, N. Bloembergen, pp. 999–1019, Columbia University Press, (1964).

(65) J. E. GEUSIC, H. M. MARCOS and L. G. VAN UITERT : ' Laser Oscillations in Nd-Doped Yttrium Aluminium, Yttrium Gallium and Gadolinium Garnets ', *Appl. Phys. Lett.*, **4**, 10, (May 15, 1964), pp. 182–184.

(66) L. F. JOHNSON and K. NASSAU : ' Infrared Fluorescence and Stimulated Emission of Nd^{3+} in $CaWO_4$ ', *Proc. IRE.*, **49**, 11, (1961), pp. 1704–1706.

(67) L. F. JOHNSON and R. A. THOMAS : ' Maser Oscillation at 0·9 and 1·35 Microns in $CaWO_4$: Nd^{3+} ', *Phys. Rev.*, **131**, 5, (Sept. 1963), pp. 2038–2040.

(68) M. I. NATHAN, W. P. DUMKE, G. BURNS, F. H. DILL and G. LASHER : ' Stimulated Emission of Radiation from GaAs *p-n* Junctions ', *Appl. Phys. Lett.*, **1**, (Nov. 1962), pp. 62–64.

(69) R. N. HALL, G. E. FENNER, J. D. KINGSLEY, T. J. SOLTYS and R. O. CARLSON : ' Coherent Light Emission from GaAs Junctions ', *Phys. Rev. Letts.*, **9**, (Nov. 1, 1962), pp. 366–367.

(70) K. WEISER and R. S. LEVITT : ' Stimulated Light Emission from Indium Phosphide ', *Appl. Phys. Lett.*, **2**, 9, (May 1, 1963), pp. 178–179.

(71) I. MELNGAILIS : ' Maser Action in InAs Diodes ', *Appl. Phys. Lett.*, **2**, 9, (May 1, 1963), pp. 176–178.

(72) R. J. PHELAN, A. R. CALAWA, R. H. REDIKER, R. J. KEYES and B. LAX : ' Infrared InSb Laser Diode in High Magnetic Fields ', *Appl. Phys. Lett.*, **3**, 9, (Nov. 1, 1963), pp. 143–145.

(73) J. F. BUTLER, A. R. CALAWA, R. J. PHELAN, T. C. HARMAN, A. J. STRAUSS and R. H. REDIKER : ' PbTe Diode Laser ', *Appl. Phys. Lett.*, **5**, 4, (Aug. 15, 1964), pp. 75–77.

(74) *IEE Jour. Quantum Elec.*, QE-1, (1965), p. 4.

(75) C. E. HURWITZ, A. R. CALAWA and R. H. REDIKER : ' Electron Beam Pumped Lasers of PbS, PbSe and PbTe ', *IEEE Jour. Quant. Elec.* QE-1, 2, (May 1965), pp. 102–103.

(76) B. S. GOLDSTEIN and J. D. WELCH : ' Microwave Modulation of a GaAs Injection Laser ', *Proc. IEEE*, **52**, 6, (1964), p. 715.

(77) W. E. AHEARN and J. W. CROWE : Intern. Quant. Elect. Conf., 1966.

(78) M. CIFTAN and P. P. DEBYE : ' On the Parameters which Affect the C.W. Output of GaAs Lasers ', *Appl. Phys. Lett.*, **6**, 6, (March 15, 1965), pp. 120–121.

(79) W. E. ENGELER and M. GARFINKEL : ' Characteristics of a Continuous High-Power GaAs Junction Laser ', *Jour. Appl. Phys.*, **35**, 6, (1964), pp. 1734–1741.

(80) J. C. MARINACE : ' High Power C. W. Operation of GaAs Injection Lasers at 77° K ', *IBM Jour. Res. & Dev.*, **8**, 5, (Nov. 1964), pp. 543–554.

(81) G. BURNS and M. I. NATHAN : ' Room Temperature Stimulated Emission ', *IBM Jour.*, **7**, (Jan. 1963), pp. 72–73.

(82) M. V. HOBDEN and M. D. STURGE : ' The Optical Absorption Edge in GaAs ', *Proc. Phys. Soc.*, A78, (1961), p. 615.

(83) M. D. STURGE : ' Optical Absorption of GaAs Between 0·6 and 2·75 eV ', *Phys. Rev.*, **127**, (1962), p. 768.

(84) J. S. FEINLEIB, W. GROVES, W. PAUL and R. ZALLEN : ' Effect of Pressure on the Spontaneous and Stimulated Emission from GaAs ', *Phys. Rev.*, **131**, 5, (1963), pp. 2070–2078.

(85) G. E. FENNER : ' Effect of Hydrostatic Pressure on the Emission from GaAs Lasers ', *Jour. Appl. Phys.*, **34**, 10, (1963), pp. 2955–2957.

(86) G. E. FENNER : ' Pressure Dependence of the Emission from $Ga(As_{1-x}P_x)$ Electroluminescent Diodes ', *Phys. Rev.*, **137**, 3A, (Feb. 1, 1965), pp. 1000–1006.

(87) M. J. STEVENSON, J. D. AXE and J. R. LANKARD : ' Line Width and Pressure Shifts in Mode Structure of Stimulated Emission from GaAs Junctions ', *IBM Jour. Res. & Dev.*, **7**, 2, (April 1963), pp. 155–156.

(88) M. PILKUHN and H. RUPPRECHT, ' Electroluminescence and Lasing Action in $GaAs_xP_{1-x}$ ', *Jour. Appl. Phys.*, **36**, 3, Part 1, (1965), pp. 684–688.

(89) I, HAYASHI, M. B. PANISH, P. W. FOY and S. SUMSKI, ' Junction Lasers which Operate Continuously at Room Temperature ', *Appl. Phys. Lett.*, **17**, 3, (Aug. 1, 1970), pp. 109–111.

(90) A. L. BLOOM, ' Gas Lasers ', *Proc. IEEE*, **54**, 10, (Oct 1966), pp. 1262–1276.

(91) W. R. BENNETT Jr., ' Inversion Mechanisms in Gas Lasers ', *Appl. Opt. Suppl.*, **2**, (1965), pp. 3–33.

(92) W. R. BENNETT JR., 'Gaseous Optical Masers', *Appl. Opt. Suppl.* **1**, *Optical Masers*, (1962), pp. 24–63.

(93) MAX BORN, ' Atomic Physics ', Blackie ; and G. HERZBERG, ' Atomic Spectra and Atomic Structure ', Dover.

(94) A. D. WHITE and J. D. RIGDEN, ' Continuous Gas Maser Operation in the Visible ', *Proc. IRE*, **50**, (July 1962), p. 1697.

(95) A. L. BLOOM, W. E. BELL and R. C. REMPEL, ' Laser Operation at $3 \cdot 39 \mu$ in a Helium-Neon Mixture, *Appl. Opt.*, **2**, 3, (March 1963), pp. 317–318.

(96) A. D. WHITE and J. D. RIGDEN, ' The effect of Superadiance at $3 \cdot 39 \mu m$ on the Visible Transitions in the He-Ne Maser ', *Appl. Phys. Lett.*, **2**, 11, (June 1 1963), pp. 211–212.

(97) A. L. BLOOM, ' Observation of New Visible Gas Laser Transitions by Removal of Dominance ', *Appl. Phys. Lett.*, **2**, 5, (March 1 1963), pp. 101–102.

(98) W. E. BELL and A. L. BLOOM, ' Zeeman Effect at $3 \cdot 39$ Microns in a Helium–Neon Laser ', *Appl. Opt.*, **3**, 3, (March 1964), pp. 413–415.

(99) E. F. LABUDA and E. I. GORDON, ' Microwave Determination of Average Electron Energy and Density in He-Ne Discharges ', *Jour. Appl. Phys.*, **35**, (May 1964), pp. 1647–1648.

(100) A. YARIV and J. P. GORDON, ' The Laser ', *Proc. IEEE*, **51**, 1, (Jan. 1963), pp. 4–23.

(101) H. A. H. BOOT, D. M. CLUNIE and R. S. A. THORN, ' Pulsed Laser Operation in a High-Pressure He-Ne Mixture ', *Nature*, **198**, 4882, (1963), pp. 773–774.

(102) G. R. FOWLES and B. D. HOPKINS, ' CW Laser Oscillation at 4416 Å in Cadmium ', *IEEE Jour. Quan. Elec.*, (Oct 1967), p. 419.

(103) W. B. BRIDGES, ' Laser Oscillation in Singly Ionized Argon in the Visible Spectrum ', *Appl. Phys. Lett.*, **4**, (April 1964), pp. 128–130.

(104) E. I. GORDON, E. F. LABUDA and W. B. BRIDGES, ' Continuous Visible Laser Action in Singly Ionized Argon, Krypton and Xenon ', *Appl. Phys. Lett.*, **4**, (May 1964), pp. 178–180.

(105) E. F. LABUDA, E. I. GORDON and R. C. MILLER, ' Continuous-Duty Argon Ion Lasers ', *IEEE Jour. Quan. Elec.*, QE-1, (September 1965), pp. 273–279.

(106) W. B. BRIDGES and A. N. CHESTER, ' Visible and UV Laser Oscillations at 118 Wavelengths in Ionized Neon and other Gases ', *Appl. Opt.*, **4**, (May 1965), pp. 573–580.

(107) P. C. CONDER and H. FOSTER, ' A Sealed-off, Beryllia Tube Argon Ion Laser ', *Proc. IERE Conf. on Lasers and Opto-Electronics*, Univ. of Southampton, March 1969.

(108) W. B. BRIDGES and A. N. CHESTER, ' Spectroscopy of Ion Lasers ' *IEEE Jour. Quan. Elec.*, QE-1, (May 1965), pp. 66–84.

(109) W. B. BRIDGES and A. S. HALSTEAD, ' New cw Laser Transitions in Argon, Krypton and Xenon ', *IEEE Jour. Quan. Elec.*, QE-2, **4**, (1966), p. 84.

(110) C. K. N. PATEL, ' Interpretation of CO_2 Optical Maser Experiments ', *Phys. Rev. Lett.*, **12**, 21, (May 25 1964), pp. 588–590.

(111) C. K. N. PATEL, ' Continuous-Wave Laser Action on Vibrational-Rotational Transitions of CO_2 ', *Phys. Rev.*, **136**, 5A (Nov. 30 1964), pp. A1187–A1193.

(112) C. K. N. PATEL, ' Selective Excitation through Vibrational Energy Transfer and Optical Maser Action in N_2-CO_2 ', *Phys. Rev. Lett.*, **13**, 21, (1964), pp. 617–619.

(113) C. P. COURTOY, *Can. Jour. Phys.*, **35**, (1957), pp. 608–648.

(114) R. L. TAYLOR and S. BITTERMAN, *Rev. Mod. Phys.*, **41**, (1961), pp. 26–47.

(115) D. C. TYTE, ' Carbon Dioxide Lasers ', *Progress in Quantum Electronics*, Academic Press, 1970.

(116) A. TRUFFERT and P. L. VAUTIER, *Onde Electrique*, **64**, (1966), pp. 417–422.

(117) C. ROSETTI, C. MEYER, P. PINSON and P. BARCHEWITZ, *C.r. hebd. Seanc. Acad. Sci.* (*B*), **264**, (1967), pp. 452–453.

(118) T. J. BRIDGES and A. R. STRAND, ' Rapid Scan Spectrometer for CO_2 Laser Studies ', *IEEE Jour. Quantum Elec.*, QE-3, (1967), pp. 336–337.

(119) C. K. N. PATEL, ' CW High Power N_2-CO_2 Laser ', *Appl. Phys. Lett.*, **7**, 1, (July 1 1965), pp. 15–17.

(120) W. J. WITTEMAN, ' Increasing Continuous Laser-Action on CO_2 Rotational Vibrational Transition through Selective Depopulation of the Lower Laser Level by Means of Water Vapour ', *Phys. Lett.*, **18**, (1965), pp. 125–127.

(121) P. A. MILES and J. W. LOTUS, ' A High-Power CO_2 Laser Radar Transmitter ', *IEEE Jour. Quan. Elec.*, QE-4, (1968), pp. 811–819.

(122) M. A. KOVACS, G. W. FLYNN and A. JAVAN, ' Q-Switching of Molecular Laser Transitions ', *Appl. Phys. Lett.*, **8**, 3, (Feb. 1 1963), pp. 62–63.

(123) G. W. FLYNN, M. A. KOVACS, C. K. RHODES and A. JAVAN, ' Vibrational and Rotational Studies Using Q-Switching of Molecular Gas Lasers ', *Appl. Phys. Lett.*, **8**, 3, (Feb. 1 1966), pp. 63–65.

(124) D. C. SMITH, ' Q-Switched CO_2 Laser ', *IEEE Jour. Quan. Elec.*, QE-5, (1969), pp. 291–292.

(125) R. DUMANCHIN and J. ROCCA-SERRA, *C.r. Acad. Sci.*, (*B*), **269**, (1970), pp. 916–917.

(126) J. BEAULIEU, *Laser Focus*, Feb. 1970, and *New Scientist*, Jan. 22 1970).

(127) D. E. MCCARTHY, ' The Reflection and Transmission of Infrared Materials. V : Spectra from 2 μ to 50 μ ', *Appl. Opt.*, **7**, (Oct. 1968), pp. 1997–2000 and ' The Reflection and Transmission of Infrared Materials. VI : Bibliography ', *Appl. Opt.*, **7**, (Nov. 1968), pp. 2221–2225.

(128) F. HORRIGAN, *Microwaves*, **8**, (1969), pp. 68–76.

(129) A. D. WHITE, ' Frequency Stabilization of Gas Lasers ', *IEEE Jour. Quan. Elec.*, QE-1, (Nov. 1965), pp. 349–357.

(130) K. D. MIELENZ, K. F. NEFFLEN, W. R. C. ROWLEY, D. C. WILSON and E. ENGELHARD, ' Reproducibility of Helium-Neon Laser Wavelengths at 633 nm ', *Appl. Oct.*, **7**, (1968), p. 289.

(131) R. L. BARGER and J. L. HALL, ' Pessure Shift and Broadening of Methane line at 3·39 μm studied by Laser-Saturated Molecular Absorption ', *Phys. Rev. Lett.*, **22**, 1, (Jan. 6 1969), pp. 4–8.

(132) C. V. RAMAN and K. S. KRISHNAN, *Nature*, **121**, (1928), p. 501.

(133) S. P. S. PORTO and D. L. WOOD, ' Ruby Optical Maser as a Raman Source ', *Appl. Opt. Supplement*, **1**, (1962), pp. 139–141.

(134) H. KOGELNIK and S. P. S. PORTO, ' Continuous He-Ne Red Laser as a Raman Source ', *Jour. Opt. Soc. Am.*, **53**, 12, (1963), pp. 1446–1447.

(135) B. SCHRADER and W. MEIER, ' Laser-Micro-Setups for Raman Spectroscopy of Fluids and Crystalline Powders ', *Z.f. Naturforschg*, **21a**, 4, (1966), pp. 480–481.

(136) E. J. WOODBURY and W. K. NG, ' Ruby Laser Operation in the near IR ', *Proc. IRE*, **50**, (1962), p. 2367.

(137) J. M. BESSON, W. PAUL and A. R. CALAWA, ' Tuning of PbSe Lasers by Hydrostatic Pressure from 8 μ to 22 μ ', *Phys. Rev.*, **173**, 3, (Sept. 15 1968), pp. 699–713.

Q

(138) R. W. Minck, R. W. Terhune and C. C. Wang, ' Nonlinear Optics ', *Appl. Opt.*, **5**, 10, (1966), pp. 1595–1612.

(139) C. C. Wang and G. W. Racette, ' Measurement of Parametric Gain Accompanying Optical Difference Frequency Generation ', *Appl. Phys. Lett.*, **6**, 8, (15 April 1965), pp. 169–171.

(140) J. A. Giordmane and R. C. Miller, ' Tunable Coherent Parametric Oscillation in LiNbo$_3$ at optical Frequencies ', *Phys. Rev. Lett.*, **14**, 27, (1965), pp. 973–976.

(141) R. G. Smith, J. E. Geusic, H. J. Levinstein, J. J. Rubin, S. Singh and L. G. Van Uitert, ' Continuous Optical Parametric Oscillation in Ba$_2$NaNb$_5$O$_{15}$ ', *Appl. Phys. Lett.*, **12**, (May 1968), p. 308.

(142) J. E. Bjorkholm, *Proc. Conf. Short Laser Pulses and Coherent Radiation.* Chania, Greece, 1969.

(143) L. B. Kreuzer, ' High Efficiency Optical Parametric Oscillation and Power Limiting in LiNbO$_3$ ', *Appl. Phys. Lett.*, **13**, 2, (July 15 1968). pp. 57–59.

(144) S. E. Harris, ' Tunable Optical Parametric Oscillators ', *Pric. IEEE,* **57**, 12, (Dec. 1969), pp. 2096–2113.

(145) J. Warner, ' Image Up-Conversion from 10·6 μm to the Visible ', *Proc. Joint Conf. on Lasers and Opto-Electronics*, University of Southampton, 1969.

(146) A. Lempicki and H. Samelson, ' Optical Maser Action in Europium Benzoylacetonate ', *Phys. Lett.*, **4**, 2, (March 15 1963), pp. 133–135.

(147) P. P. Sorokin and J. R. Lankard, ' Stimulated Emission Observed from an Organic Dye, Chloro-aluminium Pythalocyanine ', *IBM Jour. Res. Dev.*, **10**, 2, (March 1966), pp. 162–163.

(148) B. B. Snavely, ' Flashlamp-Excited Organic Dye Lasers ', *Proc. IEEE*, **57**, 8, (Aug. 1969), pp. 1374–1390.

(149) P. P. Sorokin, J. R. Lankard, E. C. Hammond and V. L. Moruzzi, ' Laser-Pumped Stimulated Emission from Organic Dyes : Experimental Studies and Analytical Comparisons ', *IBM Jour. Res. Dev.*, **11**, 2, (March 1967), pp. 130–148.

(150) P. P. Sorokin and J. R. Lanbard, ' Flashlamp Excitation of Organic Dye Lasers : A Short Communication ', *IBM Jour. Res. Dev.*, **11**, 2, (March 1967), p. 148

(151) D. J. Bradley, A. J. E. Durrant, G. M. Gale, M. Moore and P. D. Smith, ' Characteristics of Organic Dye Lasers as Tunable Frequency Sources for Nanosecond Absorption Spectroscopy ', *IEEE Jour. Quan. Elec.*, QE-4, (1968), p. 707.

(152) L. B. Kreuzer, ' Single Mode Oscillation of a Pulsed Singly Resonant Optical Parametric Oscillator ', *Appl. Phys. Lett.*, **15**, (1969), p. 263.

(153) B. H. Soffer and B. B. McFarland, ' Continuously Tunable, Narrow-Band Organic Dye Lasers ', *Appl. Phys. Lett.*, **10**, (1967), p. 266.

(154) A. Heller, ' A High-Gain Room-Temperature Liquid Laser : Trivalent Neodymium in Selenium Oxychloride ', *Appl. Phys. Lett.*, **9**, 3, (Aug. 1 1966), pp. 106–108.

(155) A. Lempicki and A. Heller, ' Characteristics of the Nd^{3+} : SeOCl$_2$ Liquid Laser ', *Appl. Phys. Lett.*, **9**, 3, (Aug. 1 1966), pp. 108–110.

(156) O. G. Peterson, S. A. Tuccio and B. B. Snavely, ' cw Operation of an Organic Dye Solution Laser ', *Appl. Phys. Lett.*, **17**, 6, (Sept. 15 1970), pp. 245–247).

(157) P. W. Harrison, ' Lasers in the Determination of Form Errors in Large Mechanical Structures ', *Proc. Conf. on Lasers and the Mechanical Engineer*, I. Mech. E., London, 1968.

(158) A. C. S. Van Heel, ' Some Practical Applications of a Precision Alignment Method ', *Appl. Scient. Res.*, B1, (1949), p. 306.

(159) J. Dyson, ' Circular and Spiral Diffraction Gratings ', *Proc. Roy. Soc.*, A 248, (1958), p. 93.

(160) H. J. Raterink, ' Applications of a c.w. Laser as a Light Source in an Optical Alignment Method ', *Z. Angew. Math. Phys.*, **16**, (1965), p. 126.

(161) W. B. Hermannsfeldt, ' Linear Alignment Techniques ', *Stanford Linear Acceleration Centre Publication PUB-82*, (March 1965).

(162) J. Dyson, ' Correction for Atmospheric Refraction in Surveying and Alignment ', *Nature*, **216**, (1967), p. 782.

(163) P. L. Bender, ' Laser Measurements of Long Distances ', *Proc. IEEE*, **55**, 6, (June 1967), pp. 1039–1045.

(164) E. R. Peck, ' Theory of the Corner-Cube Interferometer ', *Jour. Opt. Soc. Am.*, **38**, (1948), p. 1015.

(165) M. V. R. K. Murty, ' Some more Aspects of the Michelson Interferometer with Cube Corners ', *Jour. Opt. Soc. Am.*, **50**, (1960), p. 7.

(166) G. W. Stroke, ' Interferometry with Rotation-Insensitive Corner-Cube Systems and Lasers ', *Jour. Opt. Soc. Am.*, **55**, (1965), p. 330.

(167) R. Beer and D. Marjaniemi, ' Wavefronts and Construction Tolerances for a Cats-Eye Retroreflector ', *Appl. Opt.*, **5**, (1965), p. 1191.

(168) W. R. C. Rowley, ' Some Aspcets of Fringe Counting in Laser Interferometers ', *Trans. IEEE*, (IM-15), (1966), p. 146.

(169) E. R. Peck, ' Fractional Fringe Measurement with the Corner Cube Interferometer ', *Jour. Opt. Soc. Am.*, **45**, (1955), p. 795.

(170) E. I. Gordon and A. D. White, ' Single Frequency Gas Lasers at 6328 Å ', *Proc. IEEE*, **52**, 2, (1964), pp. 206–207.

(171) G. Birnbaum, ' Frequency Stabilization of Gas Lasers ', *Proc. IEEE*, **55**, (1967), p. 1015.

(172) D. M. Clunie and N. H. Rock, ' The Laser Feedback Interferometer ', *Jour. Sci. Inst.*, **41**, (1965), pp. 489–492.

(173) T. S. Taseja, A. Javan and C. H. Townes, ' Frequency Stability of He-Ne Masers and Measurements of Length ', *Phys. Rev. Lett.*, **10**, 5, (March 1 1963), pp. 165–167.

(174) E. Bergstrand, ' The Geodimeter System : A Short Discussion of its Principal Function and Future Development ', *Jour. Geophys. Res.*, **65**, (Feb. 1960), pp. 404–409.

(175) K. D. Froome and R. H. Bradsell, ' Distance Measurement by Means of Light Ray Modulated at a Microwave Frequency ', *Jour. Sci. Inst.*, **38**, (Dec. 1961), pp. 458–462.

(176) L. D. Smullin and G. Fiocco, ' Optical Echoes from the Moon ', *Nature*, **194**, (June 30 1962), p. 1267.

(177) A. Z. Grasyuk, V. S. Zuev, Y. L. Kokurin, P. G. Krynkov, V. V. Kurbasov, V. H. Lobanov, F. M. Mozhzherin, A. N. Sukhanovskii, N. S. Cherrykh, and K. K. Churaev, ' Optical Location of the Moon ', *Soviet Phys.—Dok.*, **9**, (Aug. 1964), pp. 162–163.

(178) J. E. Faller and E. J. Wampler, ' The Lunar Laser Reflector ', *Sci. Am.*, (March 1970), pp. 38–49.

(179) F. E. Birbeck and K. G. Hambleton, ' A Gallium Arsenide Laser Rangefinder Used as an Aircraft Altimeter ', *Jour. Sci. Inst.*, **42**, (Aug. 1965), pp. 541–542.

(180) R. T. H. Collis, ' Lidar : A new Atmospheric Probe ', *Quar. Jour. R. Met. Soc.*, **92**, 392, (April 1966), p. 220.

(181) R. T. H. Collins, ' Lider ', *Appl. Opt.*, **9**, 8, (Aug. 1970), pp. 1782–1788.

(182) C. A. Northend, R. C. Honey and W. E. Evans, ' Laser Radar (Lidar) for Meteorological Experiments ', *Rev. Sci. Inst.*, **37**, 4, (April 1966), p. 393.

241

(183) M. G. H. LIGDA, ' The Laser in Meteorology ', *Discovery*, (July 1965).

(184) G. FIOCCO and L. D. SMULLIN, ' Detection of Scattering Layers in the Upper Atmosphere (60–140 km) by Optical Radar ', *Nature*, **199,** (1963), p. 1275.

(185) P. M. HAMILTON, ' The use of Lidar in Air Pollution Studies ', *Air and Water Pollution*, **10,** (June 1966), p. 427.

(186) S. C. L. BOTCHERBY and G. A. BARTLEY-DENNISS, ' Length and Velocity Measurement by Laser ', *Optics Tech.*, (Feb. 1969), pp. 85–88.

(187) G. SAGNAC, *Jour. Phys. Radium*, **4,** (1914), p. 177.

(188) A. H. MICHELSON and H. G. GALE, ' The Effect of the Earth's Rotation on the Velocity of Light ', *Astro. Phys. Jour.*, **61,** 137, (April 1925), p. 140.

(189) A. ROSENTHAL, ' Regenerative Circulatory Multiple-Beam Interferometry for the Study of Light Propagation Effects ', *Jour. Opt. Soc. Am.*, **52,** (1962), p. 1143–1148.

(190) W. M. MECEK and D. T. M. DAVIS, Jr., ' Rotation Rate Sensing with Traveling-Wave Ring Lasers ', *Appl. Phys. Lett.*, **2,** (1963), p. 67.

(191) G. JOOS, ' Theoretical Physics ', Blackie, London, (1953).

(192) F. ARONOWITZ and R. J. COLLINS, ' Mode Coupling Due to Backscattering in a He-Ne Traveling-Wave Ring Lasers ', *Appl. Phys. Lett.*, **9,** (1966), p. 55.

(193) R. C. SMITH and L. S. WATKINS, ' Ring Lasers Principals and Applications ', *Proc. Conf. on Lasers and the Mechanical Engineer*, I. Mech. E., London, (1968), pp. 38–42.

(194) I. P. KAMINOW and E. H. TURNER, ' Electro-Optic Light Modulators ', *Appl. Opt.*, **5,** (1966), pp. 1612–1628.

(195) C. J. PETER, ' Gigacycle Bandwidth Coherent Light Traveling-Wave Phase Modulator ', *Proc. IEEE*, **51,** (1963), p. 147.

(196) G. GRAU and D. ROSENBERGER, ' Low Power Microwave Modulation of a 0·63 μ He-Ne Laser ', *Phys. Lett.*, **6,** (1963), p. 129.

(197) R. W. DIXON, ' Photoelastic Properties of Selected Materials and their Relevance for Applications to Acoustic Light Modulators and Scanners ', *Jour. Appl. Phys.*, **38,** (1967), pp. 5149–5153.

(198) E. G. SPENCER, P. V. LENZO and A. A. BALLMAN, ' Dielectic Materials for Electro-optic, Elasto-optic and Ultrasonic Device Applications ', *Proc. IEEE*, **55,** (1967), pp. 2074–2108.

(199) R. W. DIXON and A. N. CHESTER, ' An Acoustic Light Modulator for 10·6 μ ', *Appl. Phys. Lett.*, **9,** (1966), pp. 190–192.

(200) R. T. DENTON and T. S. KINSEL, ' Terminals for a High-Speed Optical Pulse Code Modulation Communications System ', *Proc. IEEE*, **56,** 2, (Feb. 1968), pp. 140–154.

(201) R. N. SCHWARTZ and C. H. TOWNES, ' Interstellar and Interplanetary Communications by Optical Masers ', *Nature*, **190,** (April 1961), pp. 205–208.

(202) MONTE ROSS, ' Search via Laser Receivers for Interstellar Communications ', *Proc. IEEE*, **53,** 11, (Nov. 1965),p. 1780.

(203) H. W. MOCKER and H. A. GUSTAFSON, ' New Contender for Space Communication ', *Laser Focus*, (Oct. 1970), pp. 30–33.

(204) S. E. MILLER and L. C. TILLOTSON, ' Optical Transmission Research ', *Appl. Opt.*, **5,** 10, (Oct. 1966), pp. 1538–1549.

(205) J. C. STEPHENSON, W. A. HASELTINE and C. B. MOORE, ' Atmospheric Absorption of CO_2 Laser Radiation ', *Appl. Phys. Letter*, **11,** (1967), pp. 164–166.

(206) J. C. SIMON, ' The Applications of Lasers to Telecommunications ', *Tel. Jour.*, (1965), pp. 416–421.

(207) H. A. GEBBIE, W. R. HARDING, C. HILSUM, A. W. PRYCE, and V. ROBERTS, ' Atmospheric Transmission in the 1 to 14 μ Region ', *Proc. Roy. Soc.*, **206**, (1951), p. 87.

(208) G. GOUBAU and J. R. CHRISTIAN, ' Some Aspects of Beam Waveguides for Long Distance Transmission at Optical Frequencies ', *IEEE Trans. MTT*, **12**, (1964), pp. 212–220.

(209) D. GLOGE, *Bell Syst. Tech. Jour.*, **46**, (1967), p. 721.

(210) M. PROCHAZKA *et al.*, ' Experimental Investigation of a Pipeline for Optical Communications ', *Elec. Lett.*, **3**, (1967), pp. 73–74.

(211) G. GOUBAU, ' Lenses Guide Optical Frequencies to Low-Loss Transmission ', *Elec.*, **39**, (10), (May 1966), pp. 83–89.

(212) D. W. BENEMAN, ' A Lens or Light Guide Using Convectively Distorted Thermal Gradient in Gases ', *Bell Syst. Tech. Jour.*, **43**, (1964), pp. 1469–1475.

(213) E. SNITZER and H. OSTERBERG, ' Observed Dielectric Waveguide, Modes in the Visible Spectrum ', *Jour. Opt. Soc. Am.*, **51**, (1961), pp. 499–505.

(214) A. E. KARBOWIAK, ' New Type of Waveguide for Light and Infrared Waves ', *Elec. Lett.*, **1**, (1965).

(215) K. C. KAO and G. A. HOCKHAM, ' Dielectric-fibre Surface Waveguides for Optical Frequencies ', *Proc. IEE*, **113**, (1966), pp. 1151–1158.

(216) E. R. SCHINELLER, ' Single-Mode Guide Laser Components ', *Microwaves*, **7**, (1), (1968), pp. 77–85.

(217) J. W. KLUVER, ' Laser Amplifier Noise at 3·5 Microns in Helium-Xenon ', *Jour. Appl. Phys.*, **37**, (1966), pp. 2987–2999.

(218) L. E. S. MATHIAS and N. H. ROCK, ' A Helium–Neon Laser Amplifier ', *Appl. Opt.*, **4**, (1965), pp. 133–135.

(219) P. K. CHEO and H. G. COOPER, ' Gain Characteristics of CO_2 Laser Amplifiers at 10·6 Microns ', *IEEE Jour. Quan. Elec.*, QE3, (1967), pp. 79–84.

(220) H. KOGELNIK and T. J. BRIDGES, ' A Nonresonant Multipass CO_2 Laser Amplifier ', *IEEE Jour. Quan. Elec.*, QE3, (1967), pp. 95–96.

(221) G. J. DEZENBERG and J. A. MERRITT, ' The Use of a Multipass Cell as a CO_2–N_2 Gas Laser Amplifier and Oscillator ', *Appl. Opt.*, **6**, (1967), pp. 1541–1543.

(222) J. W. CROWE and W. E. AHEARN, ' Semiconductor Laser Amplifier ', *IEEE Jour. Quan. Elec.*, QE-2, (1966), pp. 283–289.

(223) C. G. YOUNG and J. W. KANTORSKI, ' Saturation Operation and Gain Coefficient of a Neodymium Glass Amplifier ', *Appl. Opt.*, **4**, (1965), pp. 1675–1677.

(224) J. R. KERR, ' Microwave-Bandwidth Optical Receiver Systems ', *Proc. IEEE*, **55**, (1967), pp. 1686–1700.

(225) L. K. ANDERSON and B. J. MCMURTY, ' High Speed Photodetectors ', *Appl. Opt.*, **5**, (1966), pp. 1573–1587.

(226) O. E. DELANGE, ' Some Optical Communications Experiments ', *Appl. Opt.*, **9**, 5, (May 1970), pp. 1167–1175.

(227) J. LYTOLLIS, ' Optical Communication Systems—A Survey ', *Opt. Tech.*, **1**, 1, (Nov. 1968), pp. 1–10.

(228) R. F. LUCY and K. LANG, ' Optical Communications Experiments at 6328 Å and 10·6 μ ', *Appl. Opt.*, **7**, (1968), pp. 1965–1970.

(229) F. E. GOODWIN and T. A. NUSSMEIER, ' Optical Heterodyne Communications Experiments at 10·6 μ ', *IEEE Jour. Quan. Elec.*, QE-4, (1968), pp. 612–617.

(230) A. E. SEIGMAN, *Proc. IEEE*, **54**, (1966), p. 1350.

(231) O. E. DELANGE, *IEEE Spectrum*, **5**, (1968), p. 77.

(232) J. M. BENNETT and E. J. ASHLEY, ' Infrared Reflectance and Emittance of Silver and Gold Evaporated in Ultrahigh Vacuum ', *Appl. Opt.*, **4**, (Feb. 1965), pp. 221–224.

(233) R. H. FAIRBANKS and C. M. ADAMS, ' Laser Beam Fusion Welding ', *Welding Jour.*, **43**, (March 1964), pp. 975–1025.

(234) J. G. SIEKMAN and R. E. MOREJIN, ' The Mechanism of Welding with a Sealed-off Continuous CO_2 Laser ', *Philips Res. Repts.*, **23**, (1968), pp. 367–374.

(235) W. G. ALWANG, L. A. CAVANAUGH and E. SAMMARTINO, *Welding Jour.* (New York), **48**, (1969), pp. 1105–1155.

(236) M. HILLIER, ' The CO_2 Laser and its High Power Applications ', Design Engineering, (May 1969), pp. 47–49.

(237) D. L. GOODSELL, *Eng. Mat. Design.*, **12**, (1969), pp. 531–535.

(238) K. J. MILLER and J. D. NINNEKHOVEN, *Machine Design*, (Aug. 5, 1965).

(239) A. O. SCHMIDT, I. HAM and T. HOSKI, *Welding Jour. Welding Research Supplement*, No. 1965.

(240) F. P. GAGLIANO, R. M. LUMLEY and L. S. WATKINS, ' Lasers in Industry ', *Proc. IEEE*, **57**, 2, (Feb. 1969), pp. 114–147.

(241) M. I. COHEN, ' Applications of Laser Beams to Integrated Circuits ', *Elec. Equip. News*, (Jan. 1969), pp. 6–12.

(242) J. P. EPPERSON, R. W. DYER and J. C. GRZYUA, ' The Laser now a Production Tool ', *Western Electric Engr.*, **10**, (April 1966), pp. 2–9.

(243) J. C. GRZYUA and A. CHESKO, ' Laser Piercing and Reworking of Diamond Dies ', *Wire and Wire Prod.*, **41**, (Sept. 1966).

(244) S. J. LINS and R. D. MORRISON, ' Laser Induced Density Changes in Film Resistors ', *WESCON Tech. Papers*, **10**, pt. 2, (1966), paper 5.

(245) M. I. COHEN, ' Laser Beams and Integrated Circuits ', *Bell Labs. Rec.*, **45**, (Sept. 1967), pp. 247–251.

(246) B. A. UNGER and M. I. COHEN, ' Laser Trimming of Thin Film Resistors ', *Conf. Elec. Comp. Tech.*, Washington D.C., (1968).

(247) ' Resistive Film Trimming with Pulsed Argon Lasers ', Hughes Aircraft, Laser Appl. Note 3001.

(248) L. BRAUN and D. R. BREUER, ' Laser Adjustable Resistors for Precision Monolithic Circuits ', *Proc. Microelectronics Symp.*, (June 1968), St. Louis, Mo.

(249) M. I. COHEN, B. A. UNGER and J. F. MILKOWSKI, ' Laser Machining of Thin Films for Integrated Circuits ', *Bell Syst. Tech. Jour.*, **47**, (1968), pp. 385–405.

(250) R. J. MURPHY and G. J. RITTER, ' Laser Induced Etching of Metal and Semiconductor Surfaces ', *Nature*, **210**, (April 1966), pp. 191–192.

(251) R. M. LUMLEY, Annual Meeting of American Ceramic Soc., (April 23, 1968).

(252) T. J. ROWE and D. J. MOULE, ' Laser Machining of Photolithographic Masks in Thin Metallic Films ', *Proc. Symp. Lasers and the Mech. Eng. I. Mech. E.*, London, (1968), pp. 13–18.

(253) B. F. SCOTT and D. L. HODGETT, ' Pulsed Solid-State Lasers for Engineering Fabrication Processes ', *Proc. Symp. Lasers and the Mech. Eng.*, London, (1968), pp. 75–84.

(254) ' Laser Metal Removal Aids Dynamic Balancing ', *Steel*, **159**, (July 1966), pp. 28–29.

(255) ' On Line Balancing Systems ', *Laser Focus*, **4**, (June 1968), p. 3.

(256) V. G. MOSSOTTI, K. LAQUA and W. D. HAGENAK, ' Laser-Microanalysis by Atomic Absorption ', *Spectro-Chemica Acta*, **23B**, (1967), pp. 197–206.

(257) K. G. Snetsinger and K. Keil, ' Microspectro Chemical Analysisv.- Minerals with the Laser Microprobe ', *Am. Mineralogist*, **52**, (No of Dec. 1967), pp. 1842–1854.

(258) N. C. Fenner and N. R. Daly, ' An Instrument for Mass Analysis Using a Laser ', *Jour. Mat-Sci.*, **3**, (1968), pp. 259–261.

(259) M. Cane, ' CO_2 Gas Laser Speeds Carton Manufacture ', *Design Engineering*, (June 1970), pp. 65–67.

(260) F. W. Lunau and E. W. Paine, *Welding and Metal Fabrications*, **37**, (1969), pp. 9–14.

(261) A. B. J. Sullivan and P. T. Houldcraft, ' Gas–Jet Laser Cutting ', *Brit. Welding Jour.*, **14**, (1967), pp. 443–445.

(262) N. Manson, D. Smart and H. Vernon Ingram, ' Laser Ophthalmo- scope and Coherent Light ', *Brit. Jour. Ophth.*, **52**, 6, (June 1968), pp. 441–449.

(263) W. L. Makous and J. D. Gould, ' The Effect of Lasers on the Human Eye ', *IBM Jour.*, (May 1968), pp. 257–271.

(264) H. H. C. Chang and K. G. Dedrick, ' On Corneal Damage Thresholds for CO_2 Laser Radiation ', *Appl. Opt.*, **8**, (1969), pp. 826–827.

(265) N. A. Peppers, A. Vassiliades, K. G. Dedrick, H. Chang, R.R . Peabody, H. Rose and H. C. Zweng, ' Corneal Damage Thresholds for CO_2 Laser Radiation ', *Appl. Opt.*, **8**, (1969), pp. 377–381.

(266) M. J. Adams, ' Cutting and Welding Using the Carbon Dioxide Laser ', *Design Engineering*, (May 1969), pp. 47–52.

(267) E. K. Pfitzer and R. Turner, *Jour. Sci. Inst.*, (Jour. Phys. E.), Ser. 2, 1, (1968), pp. 360–361.

(268) D. Gabor, ' A New Microscopic Principle ', *Nature*, 4098, (May 15, 1948), pp. 777.

(269) D. Gabor, ' Microscopy by Reconstructed Wavefronts ', *Proc. Roy. Soc.* A197, (1949), pp. 454–487.

(270) D. Gabor, ' Microscopy by Reconstructed Wavefronts ', *Proc. Phys. Soc.*, **64**, 6, (June 1, 1951), pp. 449–469.

(271) D. B. Neumann, ' Geometrical Relationships Between the Original Object and the two Images of a Hologram Reconstruction ', *Jour. Opt. Soc. Am.*, **56**, 7, (July 1966), pp. 858–861.

(272) E. N. Leith and J. Upatnieks, ' Wavefront Reconstruction with Continuous-Tone Objects ', *Jour. Opt. Soc. Am.*, **53**, 12, (Dec. 1963), pp. 1377–1381.

(273) E. N. Leith and J. Upatnieks, ' Wavefront Reconstruction with Diffused Illumination and Three-Dimensional objects ', *Jour. Opt. Soc. Am.*, **54**, 11, (Nov. 1964), pp. 1295–1301.

(274) J. D. Riġden and E. I. Gordan, ' The Granularity of Scattered Optical Maser Light ', *Proc. IRE.*, **50**, 11, (Nov. 1962), pp. 2367–2368.

(275) L. I. Goldfischer, ' Autocorrelation Function and Power Spectral Density of Laser-Produced Speckle Patterns ', *Jour. Opt. Soc. Am.*, **55**, (1964), pp. 247–243.

(276) R. W. Meier, ' Magnification and Third-Order Aberrations in Holo- graphy ', *Jour. Opt. Soc. Am.*, **55**, 5, (Aug. 1965), pp. 987–992.

(277) G. W. Stroke, ' Lenless Fourier Transform Method for Optical Holography ', *Appl. Phys. Lett.*, **6**, (1965), p. 201.

(278) R. F. Van Ligten, ' Influence of Photographic Film on Wavefront Re- construction. I. Plane Wavefronts ', *Jour. Opt. Soc. Am.*, **56**, 1, (1966), pp. 1–9.

(279) C. C. Eaglesfield, ' Resolution of X-ray Microscopy by Hologram ', *Elec. Lett.*, **1**, 7, (Sept. 1965), pp. 181–182.

(280) R. W. MEIER, ' Cardinal Points and the Novel Imaging Properties of a Holographic System ', *Jour. Opt. Soc. Am.*, **56**, 2, (1966), pp. 219–223.

(281) E. B. CHAMPAGNE, ' Nonparaxial Imaging Magnification and Aberration, Properties in Holography ', *Jour. Opt. Soc. Am.*, **57**, (1967), pp. 51–55.

(282) I. A. ABRAMOWITZ and J. M. BALLANTYNE, ' Evaluation of Hologram Aberrations by Ray Tracing ', *Jour. Opt. Soc. Am.*, **57**, (1967), pp. 1522–1526.

(283) P. J. VAN HEERDEN, ' A New Optical Method of Storing Retaining Information ', *Appl. Opt.*, **2**, (1963), pp. 387–392 and 393–400.

(284) E. N. LEITH, A. KOZMA, J. UPATNIEKS, J. MARKS and N. MASSEY, ' Holographic Data Storage in Three-Dimensional Media ', *Appl. Opt.* **5**, (1966), pp. 1303–1311.

(285) D. GABOR and G. W. STROKE, ' The Theory of Deep Holograms ', *Proc. Roy. Soc. A*, **304**, (1968), pp. 275–289.

(286) Y. N. DENISYUK, ' Photographic Reconstruction of the Optical Properties of an Object in its own Scattered Radiation Field ', *Soviet Phys.-Doklady*, **7**, (1962), pp. 543–545.

(287) G. W. STROKE and A. E. LABEYRIE, ' White Light Reconstruction of Holographic Images Using the Lippman-Bragg Diffraction Effect ', *Phys. Lett.*, **20**, (1966), pp. 368–370.

(288) L. H. LIN, K. S. PENNINGTON, G. W. STROKE and A. E. LABEYRIE, ' Multicolour Holographic Image Reconstruction with White Light Illumination ', *Bell Syst. Tech. Jour.*, **45**, (1966), pp. 659–661.

(289) K. S. PENNINGTON and L. H. LIN, ' Multicolour Wavefront Reconstruction ', *Appl. Phys. Lett.*, **7**, (1965), pp. 56–57.

(290) A. A. FRIESEM and R. J. FEDOROWICZ, ' Multicolour Wavefront Reconstruction ', *Appl. Opt.*, **6**, 3, (March 1967), pp. 529–536.

(291) R. J. COLLIER and K. S. PENNINGTON, ' Multicolour Imaging from Holograms Formed on Two-Dimensional Media ', *Appl. Opt.*, **6**, 6, (June 1967), pp. 1091–1095.

(292) H. KOGELNIK, ' Reconstructing Response and Efficiency of Hologram Gratings ', *Proc. Symp. Modern Optics*, Polytech. Inst. Brooklyn, (March 1967), pp. 605–617.

(293) M. R. TUBBS, M. J. BEESLEY and H. FOSTER, ' Holographic Recording on Photosensitive Iodide Films ', *Brit. Jour. Appl. Phys.*, **2**, 2, (1969), pp. 197–200.

(294) N. K. SHERIDAN, ' Production of Blazed Holograms ', *Apply Phys. Lett.*, **12**, (1968), pp. 316.

(295) M. J. BEESLEY and J. G. CASTLEDINE, ' The Use of Photoresist as a Holographic Recording Medium ', *Appl. Opt.*, **9**, 12, (Dec. 1970), pp. 2620–2625.

(296) J. P. KIRK, ' Hologram on Photochromic Glass ', *Appl. Opt.*, **5**, (1966), p. 1684.

(297) G. P. SMITH, ' Photochromic Glasses : Properties and Applications ', *Jour. Mat. Sci.*, **2**, (1967), pp. 139–152.

(298) D. R. BOSOMWORTH and H. J. GERRITSEN, ' Thick Holograms in Photochromic Materials ', *Appl. Opt.*, **7**, 1, (Jan. 1968), pp. 95–98.

(299) A. L. MIKAELIANE, A. P. AXENCHIKOV, V. I. BOBRINEV, E. H. GULANIANE and V. V. SHATUN, ' Holograms on Photochromic Films ', *IEEE Jour. Quan. Elec.*, QE-4, (Nov. 1968), pp. 757–762.

(300) T. A. SHANKOFF and R. K. CURRAN, ' Efficient, High Resolution, Phase Diffraction Gratings ', *Appl. Phys. Lett.*, **13**, 7, (Oct. 1, 1968), pp. 239–241.

(301) T. A. SHANKOFF, ' Phase Holograms in Dichromated Gelatine ', *Appl. Opt.*, **7**, (1968), pp. 2101.

(302) L. H. Lin, ' Hologram Formation in Hardened Dichromated Gelatin Films ', *Appl. Opt.*, **8**, 5, (May 1969), pp. 963–966.

(303) A. Kozma, ' Photographic Recording of Spatially Modulated Coherent Light ', *Jour. Opt. Soc. Am.*, **56**, (1966), pp. 428.

(304) A. A. Friesem and J. S. Zelenka, ' Effects of Film Nonlinearities in Holography ', *Appl. Opt.*, **6**, 10, (Oct. 1967), pp. 1755–1759.

(305) J. W. Goodman and G. R. Knight, ' Effects of Film Nonlinearities on Wavefront—Reconstruction Images of Diffuse Objects ', *Jour. Opt. Soc. Am.*, **58**, 9, (Sept. 1968), pp. 1276–1283.

(306) A. Kozma, ' Analysis of the Film Non-Linearities in Hologram Recording ', *Optica Acta*, **15**, 6, (1968), pp. 527–551.

(307) A. Friesom, A. Kozma and G. F. Adams, ' Recording Parameters of Spatially Modulated Coherent Wavefronts, *Appl. Opt.*, **6**, 5, (May 1967), pp. 851–856.

(308) ' Photographic Materials for Holography ', Agfa-Geraest Technical Information (Aug. 1969).

(309) R. L. Lamberts and C. N. Kurtz, ' Bleached Holograms with Reduced Flare Light ', *Proc. Spring Meeting of Opt. Soc. Am.*, (April 1970). Also published in shortened form as a ' Reversal Bleach Process for Producing Phase Holograms on Kodak Spectroscopic Plate Type 649-F ', by Eastman Kodak, Rochester, New York.

(310) R. J. Collier, E. T. Doherty and K. S. Pennington, ' Applications of Moire-Techniques to Holography ', *Appl. Phys. Lett.*, **7**, (1965), pp. 223–225.

(311) R. E. Brooks, L. O. Heflinger and R. F. Wuerker, ' Intergerometry with a Holographically Reconstructed Comparison Beam ', *Appl. Phys. Lett.*, **7**, (1965), pp. 248–249.

(312) L. O. Heflinger, R. F. Wuerker and R. E. Brooks, ' Holographic Interferometry ', *Jour. Appl. Phys.*, **37**, 2, (Feb. 1966), pp. 642–649.

(313) J. M. Burch, A. E. Ennos ane R. J. Wilton, ' Dual- and Multiple-Beam Interferometry by Wavefront Reconstruction ', *Nature*, **209**, (1966), pp. 1015–1016.

(314) B. P. Hildebrand and K. A. Haines, ' Interferometric Measurements Using the Wave-Front Reconstruction Technique ', *Appl. Opt.*, **5**, (1966), pp. 172–173.

(315) K. A. Haines and B. P. Hildebrand, ' Surface—Deformation Measurement Using the Wavefront Reconstruction Technique ', *Appl. Opt.*, **5**, (1966), pp. 595–602.

(316) E. Archbold, J. M. Burch and A. E. Ennos, ' The Application of Holography to the Comparison of Cylinder Bores ', *Jour. Sci. Inst.*, **44**, (1967), pp. 489–494.

(317) R. L. Powell and K. A. Stetson, ' Interferometric Vibration Analysis by Wavefront Reconstruction ', *Jour. Opt. Soc. Am.*, **55**, 12, (Dec. 1965), pp. 1593–1598.

(318) K. A. Stetson and R. L. Powell, ' Interferometric Hologram Evaluation and Real-Time Vibration Analysis of Diffuse Objects ', *Jour. Opt. Soc. Am.*, **55**, (1965), pp. 1694–1695.

(319) K. A. Stetson, R. L. Powell, ' Hologram Interferometry ', *Jour. Opt. Soc. Am.*, **56**, (1966), pp. 1161–1166.

(320) E. Archbold and A. E. Ennos, ' Observation of Surface Vibration Modes by Stroboscopic Hologram Interferometry ', *Nature*, **217**, (1968), pp. 942–943.

(321) E. Archbold and A. E. Ennos, ' Techniques of Hologram Interferometry for Engineering Inspection and Vibration Analysis ', *Proc. Conf. Engineering Uses of Holography*, Strathclyde, (1968), **pp. 381–396.**

247

(322) K. HAINES and B. P. HILDEBRAND, ' Contour Generation by Wavefront Reconstruction ', *Phys. Lett.* **19**, 1, (Sept. 15, 1965), pp. 10–11.

(323) B. P. HILDEBRAND and J. A. HAINES, ' The Generation of Three-Dimensional Contour Maps by Wavefront Reconstruction ', *Phys. Lett.*, **21**, 4, (June 1, 1966), pp. 422–423.

(324) B. P. HILDEBRAND and K. A. HAINES, ' Multiple-Wavelength and Multiple-Source Holography Applied to Contour Generation ', *Jour. Opt. Soc. Am.*, **57**, 2, (Feb. 1967), pp. 155–162.

(325) GRANT, *Proc. Conf. Engineering Uses of Holography*, Strathclyde, (1968), pp. 130–131.

(326) B. A. SILVERMAN, B. J. THOMPSON and J. H. WARD, ' A Laser Fog Disdrometer ', *Journ. Appl. Met.*, **3**, (1964), pp. 792–801.

(327) B. J. THOMPSON, G. B. PARRENT, J. H. WARD and B. JUSTH, ' A Readont Technique for the Laser Fog Disdrometer ', *Jour. Appl. Met.*, **5**, (1966), pp. 343–348.

(328) B. J. THOMPSON, J. H. WARD and W. R. ZINKY, ' Applications of Holo-gram Techniques for Particle Size Analysis ', *Appl. Opt.*, **6**, (1967), pp. 519–526.

(329) B. J. THOMPSON, ' Particle Size Examination ', *Proc. Conf. Engineering Uses of Holography*, Strathclyde, (1968), pp. 249–259.

(330) W. T. WELFORD, ' Obtaining Increased Focal Depth in Bubble Chamber Photography by an Application of the Hologram Principle ', *Appl. Opt.* **5**, (1966), p. 872.

(331) J. H. WARD and B. J. THOMPSON, ' In-Line Hologram System for Bubble-Chamber Recording ', *Jour. Opt. Soc. Am.*, **57**, (1967), p. 275.

(332) R. J. WITHRINGTON, ' Bubble Chamber Holography ', *Proc. Conf. Engineering Uses of Holography*, Strathclyde, (1968), pp. 267–278.

(333) A. VANDER LUGT, F. B. ROTZ and A. KLOOSTER, JR., ' Character Reading by Optical Spatial Filtering ', *Optical and Electro-Optical Information Processing*, MIT Press, Cambridge, Mass., (1965), pp. 125–142.

(334) D. GABOR, ' Character Recognition by Holography ', *Nature*, **208**, (1965), pp. 422–423.

(335) E. N. LEITH and J. UPATNIEKS, ' Microscopy by Wavefront Reconstruc-tion ', *Jour. Opt. Soc. Am.*, **55**, 5, (May 1965), pp. 569–570.

(336) E. N. LEITH, J. UPATNIEKS and K. A. HAINES, ' Microscopy by Wave-front Reconstruction ', *Jour. Opt. Soc. Am.*, **55**, 8, (Aug. 1965), pp. 981–986.

(337) R. F. VAN LIGTEN and H. OSTERBERG, ' Holographic Microscopy ', *Nature*, **211**, (July 16, 1966), pp. 282–283.

(338) R. F. VAN LIGTEN, ' Holographic Microscopy ', *Optics Tech.*, **1**, 2, (Feb. 1969), pp. 71–77.

(339) M. J. BEESLEY, H. FOSTER and K. G. HAMBLETON, ' Holographic Projec-tion of Microcircuit Patterns ', *Elec. Lett.*, **4**, (1968), pp. 49–50.

(340) M. J. BEESLEY, ' A Potential Application of Holography to Microcircuit Manufacture ', *Proc. Conf. Engineering Uses of Holography*, Strath-clyde, (1968), pp. 503–516.

(341) H. KIEMLE, ' Holographic Micro-Images for Industrial Applications ', *Proc. Conf. Engineering Uses of Holography*, Strathclyde, (1968), pp. 517–525.

(342) G. GLATZER, ' High Resolution Holography with Large Object Fields ', *Proc. Conf. Lasers and Opto-Electronics*, Southampton, (1969).

(343) J. M. BURCH and D. A. PALMER, ' Interferometric Methods for the Photographic Production of Large Gratings ', *Optica Acta*, **8**, 1, (Jan. 1961), pp. 73–80.

(344) N. George and J. W. Matthews, ' Holographic Diffraction Gratings ', *Appl. Phys. Lett.*, **9,** 5, (Sept. 1, 1966), pp. 212–215.

(345) A. Labeyrie and J. Flamand, *Optics Comm.*, (1969), pp. 1–4.

(346) A. F. Metherell, H. M. A. El-Sum, J. J. Dreher and L. Larmore, ' Introduction to Acoustical Holography ', *Jour. Acoust. Soc. Am.*, **42,** 4, (1967), pp. 733–742.

(347) R. K. Mueller and N. K. Sheridon, ' Sound Holograms and Optical Reconstruction ', *Appl. Phys. Lett.*, **9,** 9, (Nov. 1, 1966), pp. 328–329.

(348) K. Preston, Jr. and J. L. Kreuzer, ' Ultrasound Imaging Using a Synthetic Holographic Technique', *Appl. Phys. Lett.*, **10,** (1967), p. 150.

(349) A. F. Metherell, H. M. A. El-sum, J. J. Dreher and L. Larmore, ' Optical Reconstruction from Sampled Holograms made with Sound Waves ', *Phys. Lett.*, **24A,** 10 (May 8, 1967), pp. 547–548.

(350) G. A. Massey, ' Acoustic Holography in Air with an Electronic Reference ', *Proc. IEEE.*, (June 1967), pp. 1115–1117.

(351) A. F. Metherell and H. M. A. El-Sum, ' Stimulated Reference in a Coarsely Sampled Acoustical Hologram ', *Appl. Phys. Lett.*, **11,** 1, (July 1, 1967), pp. 20–23.

(352) A. F. Metherell, ' Acoustical Holography with a Single Stationary Point Detector ', *Proc. Conf. Engineering Uses of Holography*, Strathclyde, (1968), pp. 539–546.

(353) M. D. Blue and D. Chen, ' Optical Techniques Light the Way to Mass-Storage Media ', *Electronics*, (March 3, 1969), pp. 108–113.

(354) P. J. Van Heerden, ' A New Optical Method of Storing and Retrieving Information ', *Appl. Opt.*, **2,** (1963), pp. 387–392.

(355) E. N. Leith, A. Kozma, J. Upatnieks, N. Massey and J. Marks, ' Holographic Data Storage in Three-Dimensional Media ', *Appl. Opt.*, **5,** (1966), p. 1303.

(356) L. H. Anderson, S. Brojdo, J. T. LaMacchia and L. H. Lin, ' A High-Capacity Semipermanent Optical Memory ', *IEEE. Jour. Quan. Elec.*, QE-3, **6,** (June 1967), p. 245.

(357) L. K. Anderson, ' Holographic Optical Memory for Bulk Data Storage ', *Bell. Labs. Rec.*, (Nov. 1968), pp. 319–325.

(358) A. L. Mikaeliane, V. I. Bobrinev, S. M. Naumov and L. Z. Sokolova, ' Design Principles of Holographic Memory Devices ', *IEEE Jour. Quan. Elec.*, QE-6, **4** (April 1970), pp. 193–198.

(359) D. C. J. Reid and P. Waterworth, ' Holograms for High Density Data Stores ', *Symp. Applications of Holography*, Besancon, (1970).

(360) S. E. Harris *Proc. IEEE* (Oct. 1966), p. 1401.

(361) D. J. Kuizenga and A. E. Siegman, ' FM and AM Mode Locking of the Homogonous Laser—Part 1: Theory ', *IEEE Jour. Quan. Elec.*, QE-6, (Nov. 1970), pp. 894–708.

(362) O. P. McDuff and S. E. Harris, ' Nonlinear Theory of the Internally Loss-Modulated Laser ', *IEEE Jour. Quan. Elec.*, QE-3, (March 1967), pp. 101–111.

(363) T. S. Kinsel, 'A Stabilized Mode-Locked Nd : YAG Laser Using Electronic Feedback ', *IEEE Jour. Quan. Elec.*, QE-9, (Jan. 1973), pp. 3–8.

(364) S. D. Lazenby, ' Ferranti Laser Aids to Close Air Support ', *Int. Defense Rev.*, Issue 2, 1974.

(365) *Janes Weapons Systems*, 1972–3, p. 624.

(366) *Aviation Week & Space Technology*, 3 May, 1971, p. 71.

(367) *Aviation Week & Space Technology*, 29 May, 1972, p. 14.

(368) J. Nuckolls, L. Wood, A. Thieson and G. Zimmerman, ' Laser Compression of Matter to Super-High Densities: Thermonuclear

(CTR) Applications', *Nature*, Vol. 236, No. 5368 (Sept. 15, 1972), pp. 139–142.

(369) K. BOYER, ' Power from Laser-Initiated Nuclear Fusion ', *Astronautics & Aeronautics*, Vol. 11, No. 8 (Aug. 1973), pp. 44–49.

(370) Janes Weapons Systems, 1976